PRESENCE 2

THE LANGUAGE AND THE MYSTERY OF THE UMMO PLANET DISCLOSED

Already published by the same author: http://www.denocla.com

REMEMBER TO POST
YOUR COMMENT ON AMAZON

UMMO MUSIC Band :   http://www.ummomusic.com
UMMO MUSIC, IXINAA
UMMO MUSIC, LIKE 2 OEMMIIs
UMMO MUSIC, BEST OF

Denis Roger
DENOCLA

ꝊRES'ENCE 2

THE LANGUAGE AND THE MYSTERY OF THE UMMO PLANET DISCLOSED

UMMO WORLD PUBLISHING

This book is the second book in the "Presence" series.
It is both a unique ethnological document
and a piece of in-depth linguistic research.
In actuality, this book is a paradigm shift which presents a
whole new field of work, along with amazing findings.

Discover the fascinating world of UMMO:

- The mysteries of the UMMO file disclosed
- Learn the UMMO language yourself
- A revolution for the terrestrial humanity!

www.denocla.com

Ummo documents source :
www.ummo-ciencias.org
www.ummo-sciences.org
www.denocla.com
and private collections.

Original pictures: special thanks to UMMOAELEWEE.

Digital Illustrations Davy H. —© D. R. Denocla

UMMO WORLD Publishing
8 Esp. de la Manufacture
92136 ISSY LES MOULINEAUX — France

TABLE OF CONTENT

1

Forty-Five Years of UMMO Documents

Following are the Author's answers in an interview with Morpheus magazine.

—Why are you interested in the Ummo file?

I had been interested in UFO phenomenon many years ago, but I had stopped my research when I encountered the magneto-hydro-dynamics (MHD) issue. In 1974, Prof. Auguste Meessen guessed that the magneto-hydrodynamics could be the propulsion mode for UFOs. MHD produces the Lorentz forces which can only be used in a fluid, or in an ionized gas. However it cannot be used in the intersidereal empty space, so this idea of Prof. Auguste Meessen would not be reawakened until several decades later. There were so many stumbling blocks like this that I did not do research on these issues for ten years. In 2002, one of my friends spoke to me about many odd documents he had collected. I made a connec-tion and began to study hundreds of pages of these letters. Andre-Jacques Holbecq and some of his friends were just beginning to put the website Ummo-science.org online along with all the letters they could collect. Many of the references mentioned in this book are listed at that site.

—The Ummo file contains around 1400 pages of texts currently known and 7,503 entries of "words" in the database on your web-site. In your opinion, what are the key points?

To understand this issue, I think we need to have a grid for a comprehensive global reading of both the UFO phenomenon and the Crop Circles. To promote general understanding, we also need to bear in mind a few determining events.

The first key point is that in the 1930's radio emissions were making the earth "noisy". In other words, if we did have neighbors in the cosmos, they would begin to receive these signals and to identify the source. Radio waves propagating at the speed of light would reach the stars within a radius of 10 to 15 LY in the mid 1940's during World War II. This was also the time frame of the first major contemporary UFO waves. Some elements of the Ummo case also suggest that this time frame also corresponded to a cosmological topology that allowed quick access to our planet.

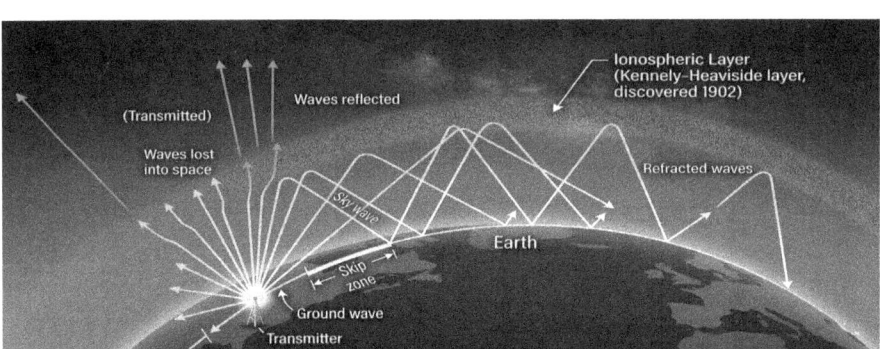

—So you think that the UFO phenomenon has intensified with the spread of radio emissions that made the earth "noisy" and with a propitious cosmological context. But if these neighbors heard us in the cosmos, as noted by the famous physicist Enrico Fermi, "Where are they?".

I think the second key point for one's overall understanding is that the reason there are varying UFO sightings is due to the fact that there are a variety of visitors, each occurring with varying frequencies and different gear, and each having different morphologies which are not easily distinguished from each other by a terrestrial observer. This explains why we usually still have a confused perception of UFO phenomenon, even though many of the stories are serious.

—Assuming that the different kinds of UFOs observed correspond to different kinds of ETs, how do you explain that there is no official manifestation from them?

If we ourselves were going to explore a distant planet inhabited by creatures less evolved than we were, we might respect a universal ethics of non-interference because any intervention could be fatal for them. . . You remember that old popular saying, "The road to Hell is paved with good intentions." Therefore, no interference. But this does not preclude reporting their presence discreetly and gradually, which is what I have called "PAX GALACTICA" in our book "PRESENCE—UFOs, Crop Circles, and Exo-Civilizations".

—If this is the case, do you think these ET people are numerous?

By merging multiple sources of information, I have come to believe that some civilizations have been regularly following our developments for several thousand years. But it is mainly since the 1950s that a hundred different civilizations have had access to our planet quite frequently.

—In your first book you show the links between the UFO phenomenon and Crops Circle which you think have a unique and very accurate origin.

Yes, in this book, I described a highly accurate hypothesis on the unique origin of the Crops Circle. According to this view they were made by E.T.s from the star Mu Herculis-A at 27.4 LY.

THE ARRIVAL OF OOMOMEN ON EARTH

—And you think that this hypothetical exocivilization coming from Mu Herculis-A is linked with the UMMO planet file?

No, not at all. But among the crowd of our curious neighbors were people from a star located at 14 LY which, according to the UMMO documents, received the 1948 radio signals from Earth. These people called themselves the phoneme "Oomomen" because the phoneme "oom-mo" means "their own planet." They are known in Spain as the "Um-mitas" which was inadvertently frenchified to "Ummites".

—What would you say are the experimental projects of these Oomomen?

According to the known documents, after several previous reconnaissance missions the Oomomen sent one of the first expeditionary ufonauts to Earth on March 28th, 1950. They are said to have landed in France in a remote area near Digne-les-Bains. According to the documents, they began to do sometimes amusing experiments on the rural French civilization at the end of the war. This contact is even more interesting because the physical similarities between Earthlings and Oomomen are amazing. Therefore, it is not really surprising that they would consider doing experiments in connection with Earthlings.

One of their experimental projects is considered to be consistent with a possible overall objective of the ETs community. It is hypothesized that this objective is to disseminate information in a progressive and controlled manner to the terrestrial populations in order to help to overcome the serious shortcomings of Earth's rulers.

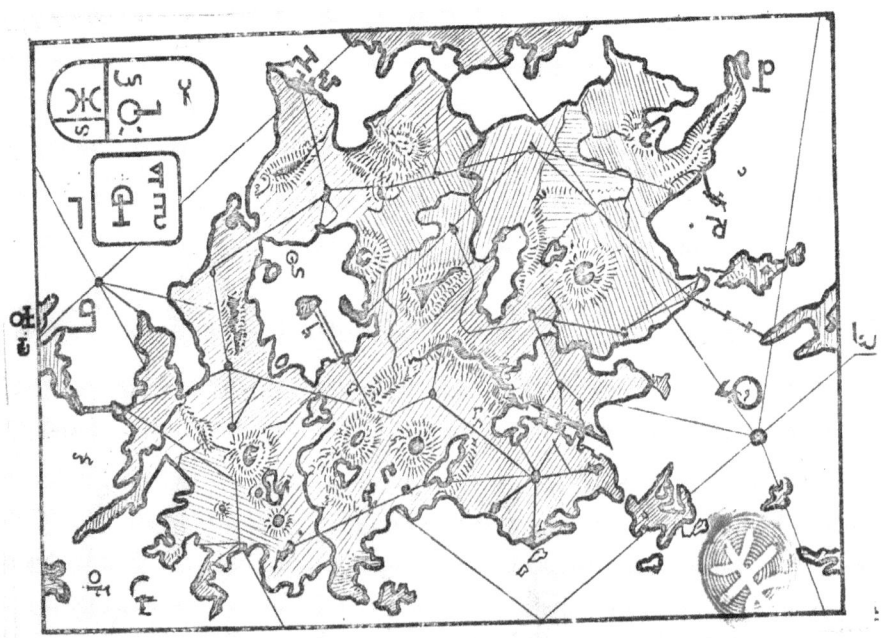

Map of the Ummo planet reconstructed from the data extracted from documents, by D. H.

THE UMMO DOCUMENTS

—But concretely speaking, how would you describe this communication experiment?

Since 1966, as part of this experiment, the Oomomen have released informational reports on various topics to many people. These documents, including the originals and copies, are known to exist. They have been preserved by their recipients in private collections. I believe that in order to allow the overall authentication of all of the documents they disseminated, the Oomomen decided to include some of the phonemes from their language in these documents. To do this they phonetically transcribed certain words into different terrestrial languages. Depending on whether the clue is clear, they translate their "words" globally, approximately, or even vaguely if the detailed information seems to be too confidential. So, I assume that the key to the authentication of the documents and of the evidence they present is in decoding the structure of the words contained in these documents. This is why we have been "discovering" these recordings since the mid-60s.

—How would this communication experiment be carried out?

Passionate about these issues, Ignacio Darnaude has done the important work of gathering information. As noted in the "Revue 2001" No. 20 published in March 1970 in Buenos Aires, an Ummo document was said to have been received on October 8th 1964 by a physics professor in Monterrey Mexico who wishes to remain anonymous. This document is known as the plot reference D612. In another document known as D108 and dated August 6th 1971 it was revealed that a team of Oomomen had worked in early 1953 in an old cottage near Marseilles on a virology research program. Then, after their first landing in June 1953, the Oomomen were said to be posing as Scandinavian veterinarians in the villa of Lady Margarita Luiz de Lihori in the small Spanish city of Albacet. Later on they took up their quarters

Denocla and Ignacio Darnaude and the Ummocat

Coffee shop named "Happy Whale"

in various geographical areas on the planet with one Oomoman responsible for each continent, for a total of 15 or 20 Oomomen.

If we reference the Ignacio Darnaude document known as D24, the communications themselves would have started later in the year 1966, fifteen years after the arrival of the first expeditionary Oomomen. The goal for our visitors was to contact people who were notoriously involved in Ufology first, and then scientists or common people found in the telephone the directory book. Regarding these extraordinary contacts the majority of the reactions were ones of rejection, and only a small group manifested themselves publicly. The group was gathered around Mr. Sesma Manzano, a ufologist who founded the "Association of Friends of Visitors from Space" in 1954. Fernando Sesma is likable, thoughtful, and easygoing, but he presents a confused mixture in his lectures which blend esoteric subjects with ufology so that many people may confuse ufology with faith and religion. Therefore the information of his lectures on the Ummo civilization is greeted with mixed reactions. Overall, they are not taken seriously by the auditors at the public meetings—which take place every Tuesday in Madrid in the Coffee shop named "Happy Whale".

Under the Francoist regime, the small amount of publicity around these occurrences limited the dissemination of information to a handful of Spaniards and a few French who also received letters in the 1970s.

—In France, who are those individuals?

To name a couple, ufologists Aimé Michel and René Fouéré received at least one letter in French sent from Berlin and referenced by Ignacio Darnaude as no. D84 dated September 04th 1969. It is available on the Ummo-science website. René Fouéré thought the letter was part of a "police operation", but Aimé

Michel was much more talkative. He spoke to a small group of polytechnic men who were investigating UFOs. Together they discovered that the letter came from a microfiche. In the 70s there was not a lot of this material available to individuals. On the other hand, one could easily assume that there might be a large number of documents on the original microfiche. In the end most of the people in this informal survey group of the French intelligentsia were literally terrified at the extent of what they felt to have discovered. So they stopped all of their investigations.

—What happened with the Ummo file in Spain?

The small isolated Spanish crew evolved rapidly and controversies erupted. In this stormy context, false documents were issued, especially by a splinter group led by Jordan Pena. These documents, pretty well-made but unable to withstand intense scrutiny, were finally identified as false. They were generally distributed along with prior authentic documents by Pena and his henchmen, the better to discredit all of the documents.

In the same way, Pena and his henchmen, including Vicente Ortuno, lit a counter-attack claiming there was false evidence of UFOs traces which were made by using buckets and torches. Meanwhile the local press "Informaciones", "Ya", and "Pueblo" reported that on February 6th 1966 at 8:00 pm a UFO landed in a field in the neighborhood of Aluche on the outskirts of Madrid and took off very quickly. The pressure calculation of the traces suggested a small machine . . . of about fifteen tons!

This event is also mentioned in a document referenced D37 sent to Fernando Sesma in February:

"In date (SPAIN) of February 6th 1966 arrived on planet EARTH three of our OAWOOLEA UEWA OEMM (spacecraft) OMWEA UMMO type [. . .] to: [. . .] SPAIN. Different people in the cities of Casilda of Bustos, Aluche (Ward C) and aviation colony witnessed our arrival. Official organizations from the Spanish capital received on February 6th and 7th a multitude of requests for explanations of the observed phenomenon, and a secret memo was sent to the Air Ministry by the flying club. The press published vague information with naive descriptions from witnesses."

The following year, Jordan Pena reiterated his counter-attack maneuvers by making fake pictures of a spacecraft similar to photographs taken by both Antonio Pardo on June 1st 1967 at 8 pm and by an anonymous person in San Jose Valderas. These photographs were published on June 2nd in the newspapers "Pueblo" and

"Informaciones" but the most noteworthy fact is that many people, such as Sesma, Villagrasa, Garrido, and Ms. Araujo, were informed of the arrival of this spacecraft on May 30th in a Oomomen's letter referenced as D60 and dated May 27th 1966. This was at least three days before the sighting and the landing.

One photograph of San Jose Valderas attributed to Antonio Pardo on June 1st 1967.

Once again, Jordan Pena was to claim that all of this was a hoax and he made all the shots. It is assumed that he did this to cover or discredit the presence of the Oomomen, either voluntarily or, more likely, because of on order from an agency like the CIA. . . This was the delight of the Ummo file "debunkers", who finally concluded that because of the forgeries the entire 1,400 pages in the Ummo file are fake.

In March 1988, Jordan Pena became severely ill from a cerebral stroke which left him very diminished and he didn't appear again for almost 5 years. Then on April 8th 1993 an article under his name appeared in the newspaper "La Alternativa Racional" claiming that he wrote the 1,400 pages of the Ummo documents by himself! So in the early 1990s for many observers the case was closed, the forger was denounced, and they were happy that the issue was concluded.

—If it is indeed an act of misinformation, what could prove or disprove the involvement of Jordan Pena in writing these letters?

Crop Circles blithely continued even after the death of the grandpa pranksters who claimed them loudly, and the Ummo documents continued to be released even after the announcement of Jordan Pena's total disability.

Because of the commotion and technical and cosmological constraints, communications were much quieter and more infrequent. However they have not ceased to this day. Indeed, the beginning of the millennium seemed to mark a new stage. The advent of the global Internet, the posting of documents claimed by the Oomomen, and the establishment of focus groups working on the subject gave rise to new letters in French. Some of them also included typical Belgian or Northern French expressions. As for the forger Jordan Pena, even though he is severely physically and mentally disabled and has been for years, he always likes to say with disdain and irony: "Believe me, I am a liar". He continues to this day to claim to be the author of the entire corpus (including new documents as soon as he learns of them) written in several languages and issued from multiple countries, for over 40 years!

—If Jordan Pena cannot have written these documents, do we know that the intelligence services could not be the perpetrators?

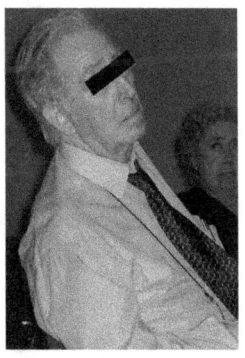

Technically, intelligence services have the means to neutralize titanic information on a massive scale. If the goals were important, they could afford an operation of this magnitude spanning a half-century. This is the "classic" hypothesis that we consider. But if we want to know who benefits from the crime, the answer is no one and nobody. Would it be in the interest of the intelligence services to create false UFO documents for half a century while they are also making every to hide UFO sightings?

Jordan Pena was severely ill by a cerebral stroke in 1988

—Except for the false documents to scramble the tracks, you categorically exclude the involvement of intelligence services in the drafting of these documents. So, what are your assumptions?

The document shipments are made for 40 years from various countries around the world: France, Spain, Britain, Canada, Malaysia, Australia, etc. They go to multiple and varied recipients depending on the subject of the document. If you assume there is a single sender, he would need to have the means and mobility to do this. If, on the contrary, there are multiple senders, the authors would be already between 60 and 70 years of age. The existing documents present subjects that are often at the forefront of knowledge at the time of the writing (For instance logic and mathematics, cosmology and physics, biology, neurology, chemistry, and the like.) As we shall see in the chapter "The mysteries of the Ummo planet file disclosed" for a single writer, this would require much more than a rare eclecticism. It presupposes that the author has had extraordinary abilities as a psychic medium, predicting 20, 25 or 35 years in advance of precise quantified results. As discussed throughout this book there is a huge and powerful cultural vision that is presented in the Ummo documents. Tetravalent logic, ontology, cosmology, primary phonetic concepts, and the language itself are gathered in a coherent, homogeneous, and indivisible set, which is what we will illustrate now...

*

The UMMO World

What is life like on the UMMO planet?

The following excerpts are taken from the actual 1500 UMMO documents, dated from 1960 to the present. The wording of these documents has been preserved exactly as translated.

The UMMO Planet

With the phoneme "XEE," we express the EIGHTEENTH part of the time interval that UMMO Planet needs to make a complete cycle moving in its orbit around the star IUMMA (you call it in your astronomical tables: WOLF 424). We're not sure whether the same star, although the characteristics and the position recorded by several ground-based observatories surprisingly coincide with our own data. But in other tables, we note serious differences regarding the star Wolf 424 (see the elements of the Yerkes Observatory which records WOLF 424 as a white dwarf nearby the Virgo constellation). So far we have not had access to terrestrial observatories to compare your records with our own tables of coordinates. You cannot have an idea of the difficulty posed by the confrontation and verification of the identity of the Stars recorded by you and us. By occupying us, another location in the Galaxy, the perspective and the relative coordinate change and that makes sense. It is also the less serious because there are conversion formulas and translational axes that make it easy. What is worse is that the estimated distances by you for many stars based on technical estimation are imprecise, so that with errors greater than 12% we can be confused with the neighbors stars. This makes it impossible to identify exactly.

How we assess the major period is different from yours and it has continued throughout our history, having its origin in an ancient astronomical measurement. We define sometimes wrongly, the XEE ("years" of UMMO) as the third revolution period of our OYAA UMMO around our sun IUMMA. The value of XEE is 77.3 Earth days.

IUMMA is a star with mass equal to 1.48 x 1030 kg. Its light spectrum is red shifted compared to your Sun with photometric indices BV and UB equal to 1.15 depending on your reference system.

UMMO revolves around IUMMA on a nearly circular trajectory eccentricity of 0.0078. The average distance UMMO IUMMA is 9.96 x 1010 meters. Another OYAA important sized NAWEE, revolves around IUMMA on an elliptical orbit of eccentricity 0.026 at an average distance of 5.97 x 1010 meters.

Ecuador	maximum radius $R = 7\,251{,}608 \times 10^3$ m
Mass of the planet	m = 9.36 x 10^{24} kg
Gravitational acceleration measured in AINNAOXOO [AINNA.OXO])	g = 11.9 m/s

Rotation on its axis: 30.92 hours (we measure in UIW [OUIW]; 30.92 h = 600 UIW [OUIW])	30.92 hours

Geological characteristics differ somewhat from those on Earth, but the atmospheric composition is very similar. (We use the Earth units in some cases). We designate the planet with a phoneme that you could translate as: UMMO. One continent and the island area occupy only 38% of the surface of our planet. UMMO moves in elliptical orbits of eccentricity 0.078 around a star called by us IUMMA (our Sun). The average distance UMMO-IUMMA is 9.96 x 1012 cm. IUMMA is a star of mass 1.48 x 1033 g. The distance between IUMMA and your Sun is around 14.42 light years. We calculate that you would localized this star at: Right ascension: 12 h 31 min; Declination: 9 ° 18 '

But the brilliance that you would notice will be greatly attenuated because of the presence of a mass of cosmic dust which attenuates and reduces it to an apparent magnitude of about 26. The surface temperature of this star is 4,580.3 degrees K. (Kelvin). Magnetic field changes are large. We record on our planet reaching 216 gauss values well above those of the Earth. These disturbances prohibit the normal use of electromagnetic frequencies, which is why we must use gravitational waves for communications.

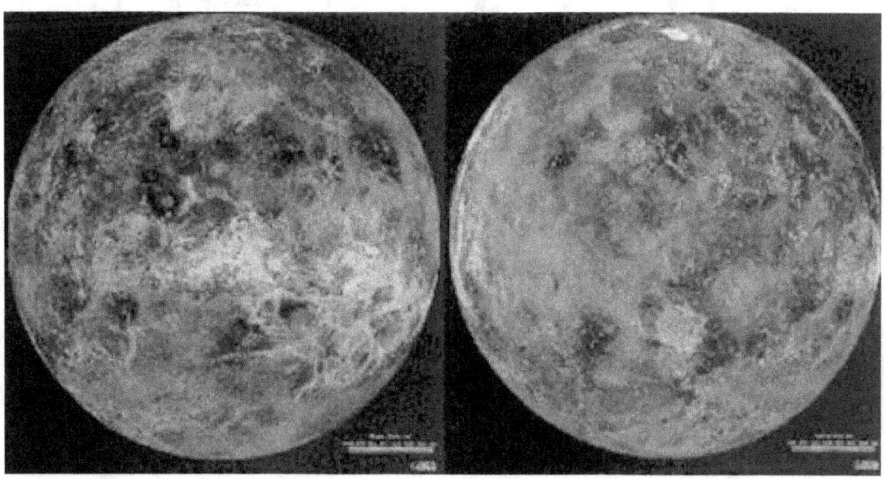

The oceans occupy approximately 62%. The rest is concentrated in one continent full of large lakes which largest one is 276,320 km² (AUWOA SAAOOA). Our mountain ranges (very eroded) are just rough shapes. The most important river of UMMOAA is the OACAWA-OEW-OEWEA 3.5 km wide at the height of UWOSS. It flows into the lake-IAWIAIA SAAOOA.

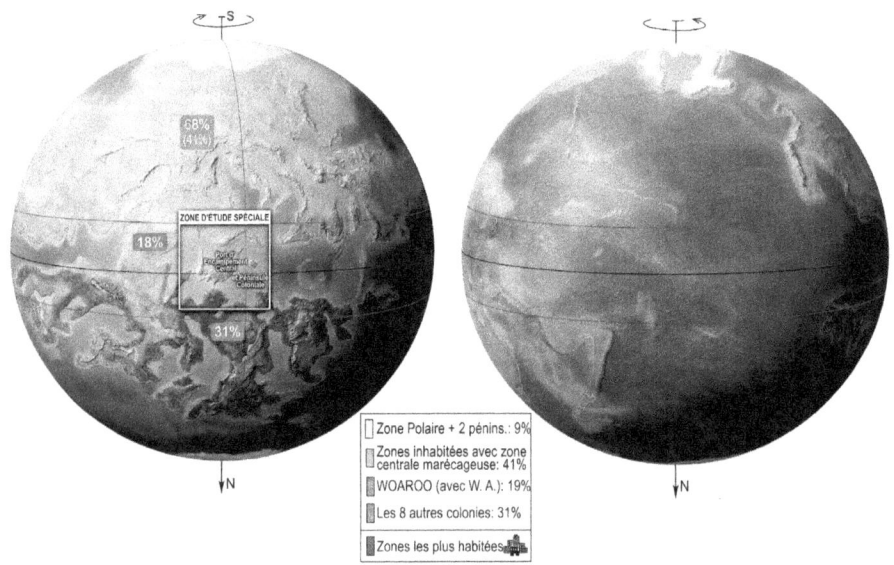

Possible UMMO World Map - Original illustration by Manuel R.

Lac AOUWOA   Bâtiment Dodécaédrique   Ile WOABAAE   Région de WOAROO AAXAA

Reconstruction based on the original map

THE GEOLOGY OF UMMO, THE OEI-OAK

The OAK-OEI are kinds of volcanoes that have the form of cracks which propel glowing columns of methane-pentane-oxygen at heights of 250 m to 6.7 km. As OASION-OEI around IA-SAAOOA Lake, its azure light illuminates the nights of UMMOAA in these regions.

Large laboratories NOOLAWE project each UIW (UMMO unit of time equal to 3.1 earth minutes) large globes of chemical compounds that come into contact with the gigantic cataracts of blue fire, sparkle like fireworks, offering a phantasmagoric spectacle. Within these globes are contained precision equipment for scientific tests.

SOME EXAMPLES OF THE UMMO'S FLORA AND FAUNA

On UMMO we have a variety of flora and fauna less rich. But in return, we enjoy an exuberant vegetation in part due to our old underground network of hydroline built with porous mortar tubes through which water is filtered, distributing it rationally in ground strata permeable, with a calculated pressure based on soil characteristics and plant species planted. Our agronomic techniques were able to provide the landscape of our planet with a rich range of shades, through the aesthetics selection and allocation of trees and shrubs gather in large forests. In our distant forest there is the tree species ANAUGAA, trees that are typical of UMMO, the IGUU and OAXAUXAA, and a vast forest of NAANAA and OBUANAA (large trees unknown on Earth, with cardioids leaves).

Illustration for approaching OBUANAA sheets

Large rocks and steep torrents. Long stems of EDIEEDI (aquatic and filamentous plant, very soft).

Illustration for approaching the long stems of EDIEEDI

Anyway, you would be surprised by the existence of OVUAANAA (trees) which sizes are near to the giant sequoias on Earth. The juicy INOWII are fruits with yellowish pulp and a brown envelope very rough.

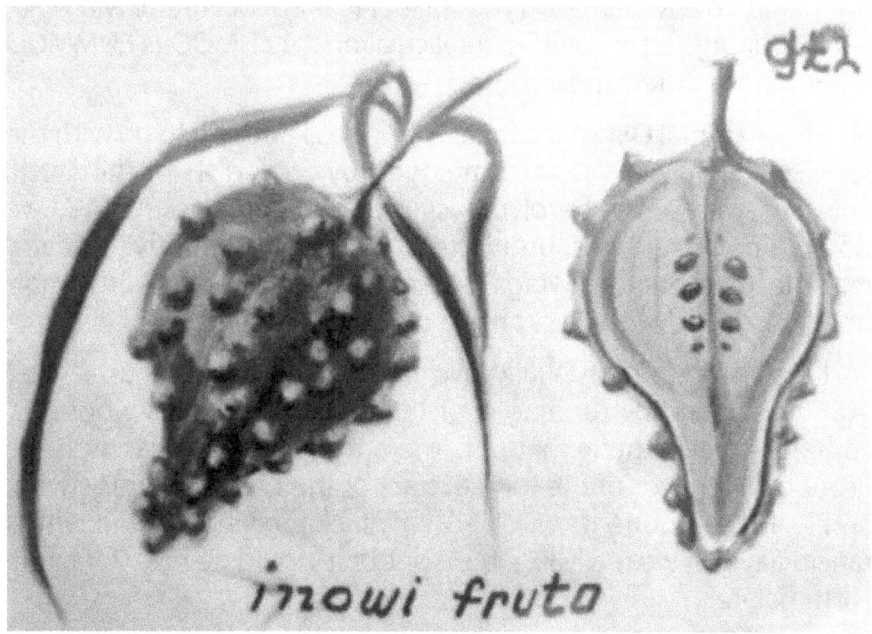

inowi fruto

Left, illustration for approaching OOGIXUA.
Right, illustration for approaching OIXIIXI.

The OOGIXUA is a vertebrate species that do not exist on Earth and has the characteristics of terrestrial REPTILES. His neck and tail are long. The four ends have a peculiar morphology that we have not found on any animal on your planet (current or antediluvian saurians). Our biggest "copies" can reach almost a ENMOO (a ENMMOO is equivalent to 1.87 meters).

The mammal OIXXIIXI whose species has nothing to do with the species known by you and cataloged by zoologists of the Earth. This large vertebrate herbivore is a flying mammal (on UMMO we do not know real birds, but there is a wide variety of flying mammals). Its membranous wings have nothing in common at the morphological level with Cheiroptera on Earth (bats).

Large amounts of OIXIIXI live in captivity for their use. Their hunting is done by running over them needles drones whose tip contains an anesthetic that puts them to sleep and to land on the ground. The flavor of the food extract of the breasts is bittersweet and its fat components will saponify quickly, giving a liquid with a rancid flavor, if we really like it, we doubt it would be tasty for some earth palate!

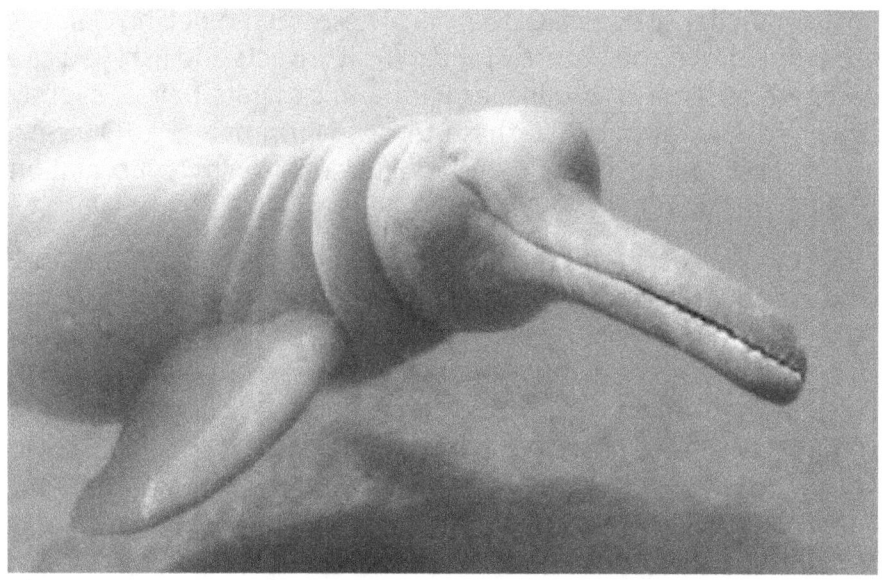

Illustration for approaching GIIDII

Other aquatic mammal, the GIIDII whose shape recalls by far the dolphin who lives in the depths of the polar seas of UMMO.

Illustration for approaching IEGOOSSAA

The animal species IEGOOSSAA. These cave-dwelling animals are omnivorous apes, mostly fond of fruit, insects and fish, you can compare to large chimpanzees with a thick white hair or slightly browned. However, unlike these terrestrial primates, IEGOOSSAA are fully bipedal and have a strong differentiation of conformation in the hands and feet.

They have no prehensile function although the toes are more developed than the vestigial of OEMMII. The IEGOOSSAA differ from other species of anthropoid from UMMO by their larger size and intelligence more acute. They have a communication language combining the complex body and facial gestures and modulated cries. They are organized into groups strongly hierarchical from which they move away only during breeding periods. After mating, all are returning to the original cell in which females give birth to their offspring that will enlarge the group. The territorial struggles are common among different groups and invariably degenerate into deadly fighting that only large adult males are involved, strongly encouraged by the cries and commotion of the females and young. The death of one of the dominant males invariably marks the end of the fight and the beaten group is savagely hunted. Individuals who died during combat are supported by the females of the victorious group and subjected to a funeral ritual in which they are covered with leaves and branches, irrespective of the original clan.

The IEGOOSSAA live at the top of our colony WOAROO that is incorporated in nature reserve. We only have relations with border groups. We trade with them voluntarily comestibles against small polished pebbles of various colors that are used primarily to our landscape art. We use them to compose frescoes, draw paths or adorn the beds of ponds and streams that we create to decorate our gardens and our collective pens. This barter avoids placing a check territorial border families to areas assigned to OEMMII and helps maintain a peaceful relationship between our two peoples.

THE DISTRIBUTION OF POPULATION AND HABITAT

On this map, the lines converge to different concentrations of population settlements directly adjacent to WOAROO AAXAA. Our eight outlying settlements represent about 31% of the total area of UMMOAA.

The inhabited area of the colony WOAROO covers about 19% in a strip of land which is approximately between the tropical upper and the equator. The polar area, including two peninsulas subpolar, is 9% of the total territory. It is cold and deserted.

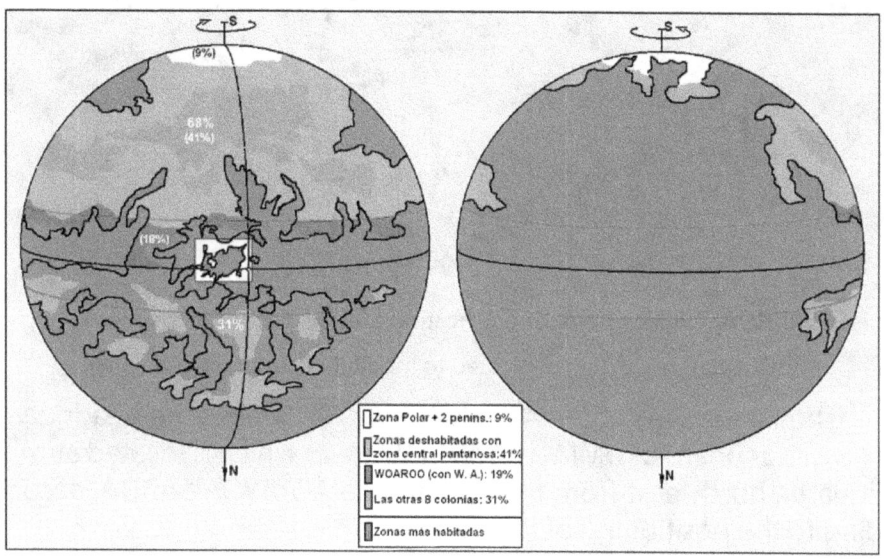

Building dodeca-hemispheric hosting officials from other planets.

Translucent egg building dedicated to meditation and religious worship.

The legislative center WOAROO AAXAA houses the headquarters of our board of UMMO (UMMOAELEWE) which is located at the equator, north-east from the largest lake AOUWOA SAAOA, according to the positioning of the card.

You may also notice, in the center of this lake, the island WOABAAAE where you can admire a large building egg-shaped translucent tan-colored, dedicated to meditation and religious worship. A splendid building dodeca-hemispherical, situated nearby, hosting officials from other planets visiting us on UMMO to maintain telepathic contact between UMMOAELEWE and peoples with whom we interact.

Our architecture is hardly comparable to yours on an aesthetic level as monumental constructions follow quite different patterns of our two worlds. We ban all angular forms especially of our constructions preferring always harmonious curves. However, just as we appreciate your pyramids, castles, temples, mosques or cathedrals, probably you would appreciate our administrative, religious or cultural buildings, often buried in the ground and stunning view outside visitors of a glazed assembly reflecting the sky, surrounded by fragrant shrubs and flower beds. On UMMO museums are dedicated to OYAGAA. Your ecological, artistic and ethnological diversity led us to build three separate buildings, each devoted to one of these three spheres and in three different colonies.

Administrative building—artist representation—Davy H.

On OOMMO, museums are dedicated to OYAGAA. Your ecological, ethnological and artistic diversity has led us to build three separate buildings, each dedicated to one of these three spheres and located in three different colonies.

THE CLIMATE OF UMMO: AN ETERNAL COOL SPRING

The climate of the planet UMMO has characteristics very different from those of Earth, creating an environment that is both harsh yet very stable. According to the document, here are the major factors that define the climate and atmosphere of this planet:

The star IUMMA (their "Sun"): The planet's climate is directly dictated by its star, IUMMA. It is a star much less luminous than our Sun. It has about 75% of our Sun's mass, and its luminosity is only about 15% to 20% of that of our star.

Absence of distinct seasons: UMMO's axis of rotation is inclined by only 18° 40' relative to the ecliptic (compared to 23° 27' for Earth). Because of this slight inclination, the seasons on UMMO are barely noticeable. In the tropical and equatorial regions (where most of the population lives), the seasons are even considered undetectable.

A "Perpetual Cool Spring": Overall, in inhabited areas, the climate resembles that of northern Finland in late September. There are no scorching summers or extremely frigid winters, but rather a "perpetual cool spring" that fosters heavily forested landscapes. Temperatures vary slightly throughout the year, with extremes estimated between +20°C at the warmest and -5°C at the coldest.

Long and very cold nights: The day and night cycles on UMMO are longer than on Earth. Furthermore, nights there are particularly cold, even when the star IUMMA illuminates the equatorial region almost perpendicularly. For example, outside a XAABI (the typical dwelling), temperature readings at dawn can drop to around 3°C.

Sky colors: The atmosphere and the nature of the star IUMMA create unique celestial hues. On the horizon, during dawn or dusk, the sky takes on a soft glow blending indigo and reddish tones.

In summary, the inhabitants of UMMO live under the gentle light of a distant sun, in a heavily forested world where temperatures are consistently cool and nights are freezing, never experiencing the stifling heat of our Earthly summers.

THE UMMO DWELLING, THE XAABII

All our humanity doesn't live in these homes scattered in the countryside. Approximately 27% of the population lives, for professional reasons, in large colonies or cities that are somewhat akin to urban-land gardens. These XAABII towers emerge few times, especially at night, from their pit-shaped wells in which they can move up or down at will. For a terrestrial observer, the UMMO countryside at nightfall seems filled with terrestrial coastal lighthouses. This is because our homes are roundabouts at the will of their owners to enable them to have the vision continue to the horizon. The lights then describe a rotational motion that is causing this little optical illusion.

The UMMO dwelling, the XAABII

With miniaturized equipment incorporated in our throat we can express ourselves with sounds perfectly intelligible. Despite this, the frequency range is more restricted than on earth (for terrestrials, it is approximately between 20 and 10,000 cycles per second). With an artificial way, we may emit through a simple conversion of harmonic frequencies as 15,000 and 20,000 cycles per second (ULTRASOUND) properly coded. Any room can be converted into a bedroom, a "meditarium", a kitchen or a playroom. If in each of our tower exist five or six (usually six) of these IAXAABI or rooms, it is because that at some point, one of which may be used for example as a bedroom for the children while in the second the father makes the meal and while in the third, converted into a bathroom, the wife or YIE blend fuel for the steam bath which precedes the meal in the afternoon. An ultrasonic coded cry activates the mechanisms that turn on the various home furnishings. An ultrasonic acoustic signal that corresponds precisely to the XAXOOU (SEAT) at his feet and opens automatically. In picture 6, you can read about the system we use in our homes for us to sit down. A small pit used to put our legs.

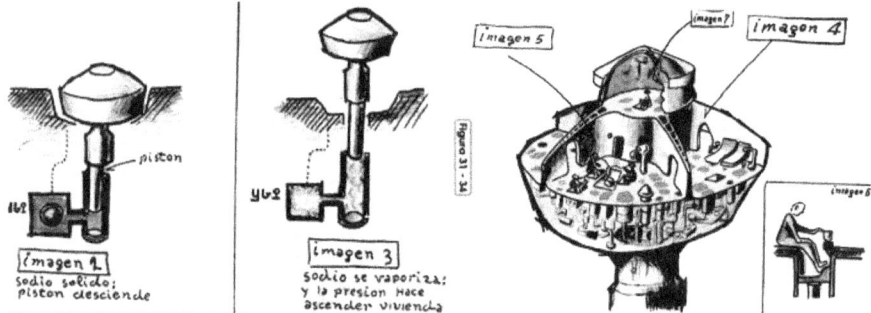

pictures: 2 – 3 – 4 – 5 – 6 - 7

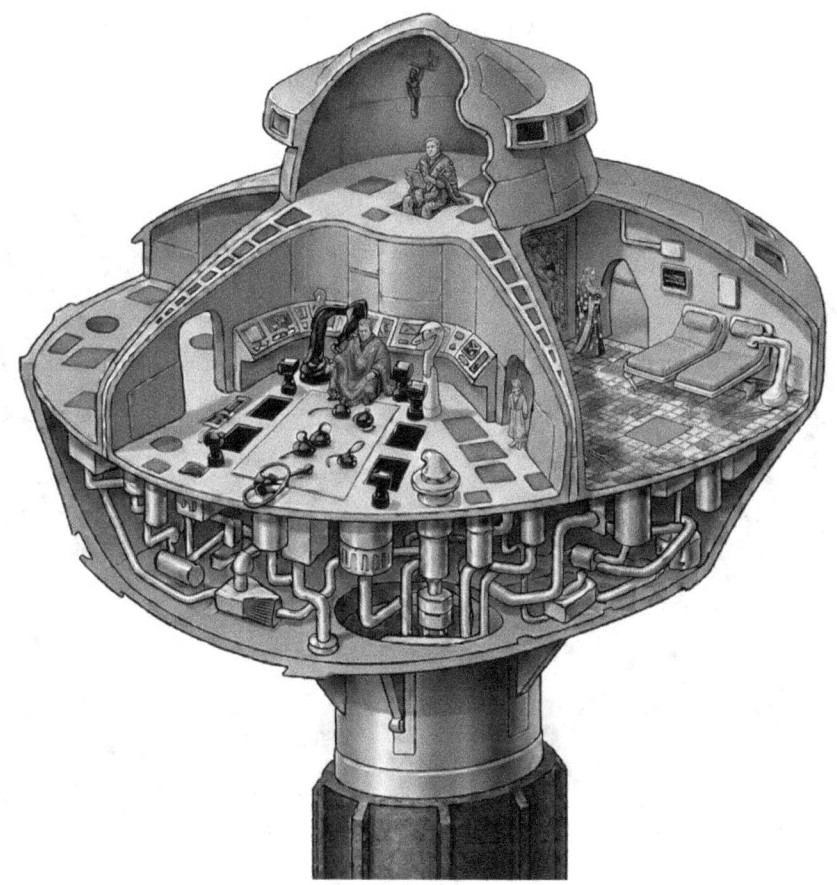

Ummo Construction, a type of XAABII

A NIGHT IN A XAABII

Our nights are intensely dark. We cannot enjoy as the Earthlings this wonderful show such as the lunar satellite (that you forget, submerged in these concrete monsters, asphalt and steel of large cities). The stars that we observe appear as sharp when the sky is clear of cloud concentrations. By cons, what we call UULibooa, (similar to polar terrestrial aurora), are much more frequent than on Earth planet, even in equatorial latitudes.

So, the sky takes an impressive appearance. Long ribbons or green and purple bands appear suspended at different heights (the higher the latitude, the higher bands have a higher vertical). On the horizon, green or magenta chromatics becomes pale yellow or slightly orange. These UULIBOOA are common in activity periods of our solar star IUMMAA. Then it is difficult to sleep, especially if we are children, contemplating the magnificent spectacle above our heads!

UULibooa

We say that it is 196 uiw.* On the horizon we begin to see a soft radiance between indigo and reddish due to the appearance of IUMMAA. Out of the XAABI, there is a temperature equal to 3 ° Celsius. Both spouses in a room in the dark begin the very brief task to undress, taking immediate antiseptic action of personal hygiene through an ablution by an adequate solution that is rarely fragrant and spraying and irradiation controlled on the eyes, mouth and nostrils.

The ablution is performed in a turbulent mass of water which is renewed at high speed with various degrees of dissolution, temperature and ultrasonic vibration. For this, each person is fully introduced in the XAXOOU (CHAIR) whose lower deck goes down, and then the pit is filled with liquid while a transparent cover made of

_____
* About 8 hours.

a semi-fluid plate, which solidifies rapidly, protects the tile of the XAABI from splashing.

fig. 4—WOIOA

Once the couple is sitting (GEE and YIE), once the IAXAABI lit with a dim light cyan, take place the UIW 24 (1 UIW = 3.1 minutes) that the OMGEEYIE devoted to meditation and prayer preceding sleep. New acoustic signals will close the seats. The facing of the IAXAABII shine with a soft magenta or purple color. Purple, green, blue, cyan and purple, with low light, are the colors selected. The couple then turns off the light to undress them both, then they light up again and will start two WOIOA devices (see Figure 4), you would call beds seeing this drawing schematically drawn although their structure look like a terrestrial couch. It is therefore necessary to describe the WOIOA.

Two discs emerge from the soil and are rapidly separated there from by a system of electromagnetic levitation. (You know this principle although its realization is still expensive). (A high frequency field can keep hanging in the air any toroïdal metal ring). At the level of these rings begin to form a conglomeration of chemical-organic foam that solidifies quickly. A gaseous system management process requires this sparkling and amorphous agglomerate to take a form of spongy sofa.

Our mating type relations are developing on a plane different of Earth. First, sexual psychology differs from those of OEMII (men and women) of this planet. Education is developed by different standards and finally the practices and habits adopted have specific differences, of course. We will try to summarize this complex matter. The sexual evolution develops through physiological and psychological processes that show marked differences. First, puberty begins in UGEE and UUYIE (children) around fourteen terrestrial years, the first menstrual period occurs between 15.5 and 16.6 years for girls. An important factor to consider is the location of epidermal areas which are associated with an erogenous marked tone. For the YIE (woman) such erogenous zones are located in the external reproductive organs, stomach and buttocks, breasts and hips. The greatest insensitivity is in the facial areas (not even on the lips), thighs and throat.

For the GEE (Male) for localized trends in female libido is directed on the reproductive organs, stomach and buttocks. Do not be surprised that the kiss is totally lacking in sexual meaning for us. This practice only known on Earth could be considered to us as repulsive without erotic purpose, we judge it as unhygienic and sterile. The beginning of the orgasm is achieved generally by touch with hands. The caress directed onto erogenous zones provokes in us an emotional-sexual effect much greater than that obtained by you. The technique of coitus has a great analogy with Western practices of Earth planet.

We did not know and we proved the futility of certain forms recommended by the Japanese and Hindu philosophy. However, two differences can be reported: defloration is never done by someone other than the husband. There is no technical hymenectomie with surgical practice, conducted by specialists of "medicine". The measure of pleasure (not only sexual) we do it through the evaluation of a function, reflected by a range of electrostatic fields generated by the cerebral cortex. This allows us to show you as curious that the duration of orgasm and its relative intensity is more balanced between men and women of UMMO as between Earthlings for which the period is more pronounced for females. Sexual psychology is of importance to us as significant as for the people of Earth. But all that is related to reproductive function is completely without prejudice or without, as you call them, "TABOOS".

Sex education for our UUGEEYIE (boys and girls) begins at a younger age than that which corresponds to the edge of adolescence. In addition training is imbued with a deep religious feeling.

Personal hygiene by washing is not recommended either within 600 UIW (30 hours) before the act, to allow a normal epidermal exudation for both spouses. This factor is very important for us for sex as our sense of smell, usually very sensitive, traditionally requires to be stimulated by the particular components of the sweat of our spouse. To the point that in the past, when was composed of a sequence of flavors for this act, we specifically studied samples of sweat of the man and woman so that the olfactory composition enjoys a relationship a syntonic and aesthetic with them. When the first sexual act will come true, we cannot, strictly speaking about defloration or ruptured hymen as our YIE (women) do not have a hymen with similar characteristics to yours. The technique of coitus between us is more uniform, less rich in terms than humans of Earth. Some ways to copulate, which in past times were used in different regions of our planet, have been abandoned by us. Contributing to this, advice and psychic anatomical characteristics instilled in successive generations by our specialists.

We reject the physical custom you use to place the woman under the man. Both begin the act by exploring mutually with the sensors of our fingers. The tips of our fingers have, unlike those of humans of Earth, a very special histological structure in the skin, similar in some aspects to the retina of the eyeball. Neuronal photoreceptors mingling with other pressure and temperature sensors, so sensitive that move the fingers at a distance of ten millimeters or more on the surface of any size, smooth or rough, colored or black, exudate or not, produces a combination of sensations that we suspect to be of a different nature from that experienced by you.

For several UIW continues this game of love exploratory in which both hands barely graze or caress the skin to finish being stimulated the same way as that experienced by you. The oral caress of the reproductive organs is more common between us than you and was also condemned as dangerous for certain areas of the cerebellum, it was an ancient practice to get into a bridal chamber maintained in a rotating special regime of accelerations and decelerations roundabouts that stimulated orgasm in us. In the terrestrial elevators in buildings with large velocities requiring high accelerations at arrival points (elevators in high buildings), we are frankly embarrassed to control our gestures of pleasure that may look like the unconscious mimicry during the sexual orgasm.

Otherwise the process has similar characteristics if we except something that is more usual for you. Control the moment of orgasm and ejaculation is greater. A sensor connected by gravitational waves in XANMOOAIUBAA records the time of ejaculation and hence the consumption of a legal union. Though, the orgasm is very slow. Insemination performed, the penis is withdrawn at the moment and fingers of the man continue exploring the vagina. Note that the only one of the few inhabited planets that we know where the society was so upset that it tolerates large intervals between puberty and rational sexual union is Earth. We do not know endemic diseases such as prostitution or various sexual perversions.

In the morning at home a rigid standard of discipline sets the time of awakening. Adults and children exchange greetings with each other, if they sleep in the same WOIWOIXAABI (bedroom) they immediately cover themselves with GIUDUUDAA EEWE (sort of porous cloak, extended it to the shape of a circle with a central hole which one introduces is head and two smaller holes for the arms).

Poncho

The garment fits the nature of his work. This is particularly the case in which we are commenting, the UULWA AGIADAA EEWE, kind of tight fitting overalls, which in this case the colors, yellow circles on a purple background, is a complex code of colors and geometric shapes representing different professional specialties of our world. The name of the garment that is associated with the profession.

The mother remains in a room that will turn into EXAABI (you can translate into the bathroom). She brings out from the ground a flexible snaking tube with a head that has a large disk with a multitude of nozzles and buttons. She manipulates them by spraying out of the mouths of a multitude of jets sprayed perfume. She opens and closes some others, sniffing with great attention the mixture.

The IAI KEAI (ART OF MIXING AROMATIC SPECIES) is a very old practice of UMMO. The YIE have always been superior to men in this difficult art. The woman devoted to this careful work almost 7 UIW (some 20 minutes) by paying as much interest as Earthly bride seasoning dishes. Adult members of the family alternated each day to this work. It is not difficult to guess, especially children and even other the brands of celebration or mood, poorly contained when it was the turn of the wife to mix species. When a woman stands out in this very difficult ancient art, the UMMOAELEWE distinctions awarded her and asked her to become a counselor or teacher in UNAWO IU (University). Our sense of smell is much more developed than yours. The difference of level is similar but in your favor in terms of music, an area where terrestrials are wonderful masters. In ancient times, the rich range of IAI (perfume) was extracted from aromatic plants, shrubs and some animal species resembling the terrestrial mollusks. Today, the variety has been greatly enriched and all flavors are produced in part synthetically.

We say partly because we have brothers who prefer, as a rite, continue to select plants in forests and then distill them in their own homes. It's a hobby like in your case the collection of postage stamps or small carved ivory statues. When the operator has found a satisfying blend, she does not forget to record the dosage, appropriately codified, in his little AARBI OMAIU (this unit is the equivalent of a terrestrial tape recording but does not tape by magnetic there is a built-in memory of a TITANIUM crystal chemically pure). However, a woman who claims to be a good IAI YIEKEAI (mixing of species) will try to never repeat the same mixture except if it was exceptionally successful.

Our eidetic olfactory (odor memory) is very important and we can remember a known mixture, as you identify a painting by attributing it to this famous artist. Finally, our sister's example that we expose, has achieved a blend that she considers as nice. In another room, everyone is waiting for the beginning of the bath. At the call of the wife, parents, spouse and child come running. They get inside the EXAABI and quickly flush from the soil panels similar to Japanese screens. Then they all undress. It is incorrect to see the nakedness of others even if they are of the same sex. Meanwhile, the atmosphere of the room is full flavored steam, the steam with a large amount of O3 (Ozone) and a multitude of aromatic components, are the first swim of the day. Everybody laughs and talks incessantly through the thin screens and then the sweet flavor of the sequences will decrease. Remember that blends succeed throughout the session, as a symphony of olfactory nuances.

On exceptional occasions and in popular concentrations comparable to those you hold on planet Earth for your sporting events, taking place on UMMO authentic IAIKEAIUUXAA (shows perfume blends) in which the aromatic nuances succeed in an exhilarating atmosphere obeying to stringent aesthetic standards.

A MEAL ON UMMO

Today they will take this meal inside the XAABIUANAA (house) but the day before it happened between the rocks along the creek near the grove of high IGUU (plant with a certain resemblance to the ancient tree ferns of the Earth Planet).

Our food: natural, artificial or semi-developed, are developed in industrial plants and underground culture in many KOAE of distance (1 Km KOAE = 8.7) and stored in cylinders of equal diameter. So the juicy INOWII—see image 8—(fruit with yellowish pulp and brown envelopes very rough) are first treated by neutron bombardment to remove any germs that may cause their spoilage. A fruit thus prepared retains its aroma, flavor and other qualities, even if not kept in a refrigerated environment and can withstand several months without deterioration.

Image 8—INOWII

But to avoid dehydration, (that is to say it dries) the INOWII are packed with a gelatinous substance (a colored compound of silicon), the assembly having a cylindrical shape. Fruits remain wrapped in this semi-transparent viscous and pinkish substance. The outer layer quickly hardens on contact with nitrogen, so that would recall you the appearance of a jar of fruit in syrup.

But in your case, there is no container of glass and gelatin is not edible, it is only a protective product that should not be exposed to air. Fats, meats, plants, glucose (sugar) are also prepared in the form of cylinders of equal size (standard on all UMMO), in a manner similar to the one you use for conservation in metal containers (cans). Similarly, other non-food consumption but topical in the homes of our planet such as acids, liquid propane, silicone polymers, zirconium oxide, germanium compounds are compressed into a cylinder (if they are powdery or solids) or poured into cylindrical containers if they are liquid. Distributing centers exists on UMMO linked by an extensive network of underground pipes NUUDAIAA or conduction at all XAABII (HOUSES) in the region.

These products (food or not) are led in cylinders NUUGII along the pipeline, with similar shape to that you employed in pneumatic tubes on Earth. The difference is that these NUUDAIAA not use pressurized air as your systems, but the HELIUM (inert gas) for the simple reason that oxygen etching the gelatinous substance that envelops much of NUUGII (receptacles for traveling). The inhabitants of the house can count on a good supply of consumer products that automatically renews itself, stored in the tubes NUUYAA and arriving periodically by pneumatic lines from supply centers. Much like a can stock.

What happens if at some point people decide to eat the sweet fruits of INOWII? The process is quite complex, but in practice it is realized in less than a tenth of UIW. A selector ring (not drawn on the image that we send you) automatically selected by rotating the cylinder NUUGII that keeps the fruit. This one is taken inside the ANAUANAA (axis of the house) to the room where the fruit has been requested. But before you go outside, it must pass through auxiliary equipment consists of an enclosure where the NUUGII (conservative gelatin cylinder) is subjected to the action of liquid oxygen.

This one attack strongly the protective gelatinous substance and dissolves it. Residues (silicon oxide and polyvinyl) are eliminated. The process should take place quickly to prevent the low temperature of liquid oxygen (185 degrees Celsius below zero) destroys the cells of the fruit. The INOWII is now freed from its protective coating and ready to be ingested (if our brother does not want to chew the fruit that came out very cold in the process), it passes through a cylinder induction, which slightly increases the temperature in an instant. The juicy INOWII is again in our hands, fresh sour, as tasty and aromatic than when he was plucked from the tree planted in large underground where artificial culture allows many annual crops.

Let our brother trying to manipulate his UAMMIXANMOO. This unit whose very detailed technical description—we see no harm in sending you in a forthcoming report (if you're interested)—has a TITANIUM memory capable of decoding a sequence of complex programs with reaped the instructions to automatically prepare that you call dishes or stews.

In principle, this device could perform all operations without the help of the operator. TITANIUM memory dictates a series of "routines and subroutines" (in computers Earthling language) to the servo unit. We will illustrate this with an example:

- Ask the YAA (tubular stores already described) a NUUGII MEAT of OOGIXUAA (popular reptile white meat).

- Eliminating the protective gelatin with liquid oxygen.

- Subjecting meat to a processing of removing toxins.

- Marinate the meat or cut into pieces (depending on the program).

- All seasoned with WUUNUA (condiment tonic).

. Add IDIA OIXII (milk fat OIXIIXI of flying mammals).

Finally, all through a box (you would call OVEN) in which, even if the temperature is the same as that of the environment, the meat is heated, causing molecular agitation of its tissues. Meanwhile, the parents and youngest son were again dressed with other showy EEWEE (tunics circular) of rich colors (those that they themselves have used previously were thrown in the IMAAUIII (species gully) and decomposed by the action of acids and then disintegrated as garbage, to be finally converted to hydrogen. Our clothes are almost never used and washed several times. They are made some UIW before being used in the XAABII (house). We do not "know" tissues.

The first meal of the day will begin. Dawn has just ended and our radiant IUMMA stands between large orange and indigo mists of the morning. You might think that this step corresponds to the breakfast, really is the case, however, with the difference that it is the longest and most plentiful of the day. That is to say the one with the richest caloric content at this time interval of 600 UIW.

Imagen 11

Imagen 10

Pictures 10 - 11—XAXOOU

The UGEE (CHILD) also make emerging from the tiled flexible metal tube (picture 11) where appears at its end a filter spray. A cloud of fine droplets yellowish is sprayed onto the ground.

You might think we're going to paint the floor. In reality, this is to create a thin film that protects the floor and serves as a tablecloth.

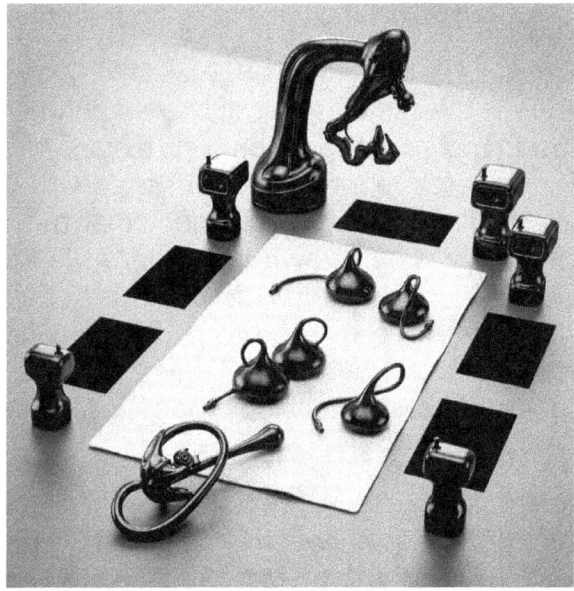

From the ground raises equipment that we call NAAXUNII which we will explain the function further. The GEE and YIE (husband and wife) put on the ground IOAOOI (CONTAINERS FOR LIQUID FOOD). Each IOAOOI consists of 3 or 4 spherical containers or enclosures that maintain until their intake broths cooked at a constant temperature.

Iaooi

We call these broths UAMIIGODAA. They are sucked by long flexible tubes.

The UOUAMII (MEAL NUMBER ZERO) will begin. It starts with some silent UIW. All close their eyes and try to accommodate their minds with a constellation of pleasant memories. Various memories, mood traits, incidents of celebrations. This preamble is necessary to impregnate the course of the meal with an atmosphere of peace and safe joy.

Everybody put in their mouths the ends (you say filters) of these long tubes which leave from IOAOOI. This will remind you of hookah oriental smoking on Earth. Aspiration of UAMIIGOODAA (liquid food) is started. The variety of these is very large. These mixtures of plants, animal extracts, synthetic aromatic essences, etc.

Compositions that make you probably think of soups or exotic sauces from distant lands on Earth. These are high in lipids or fat, with a high percentage of carbohydrates tonic and flavored containing a wide range calibrated (for each individual) factors necessary for human metabolism (glucose, galactose, fatty acids, chlorides, calcium, potassium, thiamine, riboflavin, nicotinamide, ascorbic acid. . .).

All are known by you on Earth (except for some compounds of phosphorus and manganese). Some of these UAMIIGOODAA are slightly sweetish; others have a sour taste flavored and slightly salty for the last ones. We can compare them with some terrestrial foods and remember their tastes similar for example to certain shellfish and edible seaweed very popular in Japan.

The timing of UAMIIGOOINUU (SOLID FOOD) arrived, but before any subsequent users of the NAAXUNII. This requires a brief explanation: On our planet, the food is placed in the mouth when they are solid, using the fingers. These have now acquired a great agility to take them. As it is considered unhygienic and unpleasant that the epidermis is stained or attacked by the greasy components, dyes, etc.

In ancient times, were used transparent gloves for this function. But long ago, at mealtimes, in the Xaabiuanaa (HOUSE) there is this device or equipment where our brothers introduce hands (NAAXUNII).

Noaxunii

Equipment where our brothers introduce hands.

These are sprayed with a fine spray (hydrosols dosed) that will solidify in contact with skin and wrap it with a thin protective layer that rejects (by surface tension effects) all greasy substances. Whenever we change course (as you say) we can in the same NAAXUNII dissolve this slim glove similar to artificial collodion used by chemists on Earth, and replace it with another, which is equivalent to wash our hands without using water or detergents. This does not mean that washing does not exist, but we'll talk further.

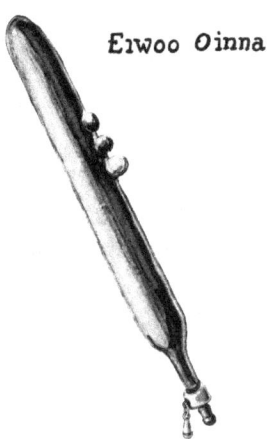

Eıwoo Oinna

We do not use knives. Our EIWOO OINNA performs the technical functions of "cutter". Its outer shape recalled one of these pencils multiple used by Earthlings. It emits at its end a thin high-energy beams wave (constant frequency of 7.9×1014 cycles per second) that cuts cleanly foods to a depth that can be adjusted by a focusing of the beam high frequency cone (this frequency is within the range that you call ultraviolet).

A meal, by D.H.

This second stage of the meal begins with slices of AIMMOA that are eaten separately from other foods. AIMMOA is a large fruit with a sweet flavor, a cellular structure pulp and a composition rich in starchy carbohydrates, no fat and has been on our planet ever since very ancient times, the staple of the UMMOAOO (UMMO culture) as for you the bread or the fruit of the ARTHOCARPUS in Ceylon or in Polynesia.

Illustration for approaching AIMMOA

Once the heart or the endocarp of the AIMMOA is eliminated, it is cut into large annular washers and eaten alone or impregnated with OIBIIA (extract of a marine animal fat) or sprayed with pure maltose or gasified with some synthetic species.

The guests are introducing again in their hands in the NAAXUNII to dissolve the protective layer and renew it.

The following food can be the tasty meat of OOGIXUAA. Meat amber (yellowish white). Its meat is the most popular on UMMO, and we are sure that the greatest palate on Earth, if they were without prejudice onto the origin of such feed, would find it very tasty. On our planet, we do not use for food processing plants or meat you call fried, any type of animal or vegetable fat. Cooking takes place through various conventional types of animal milk, the IDIIA. Two of them are very popular with rich content of fatty acids. One is extracted from mammalian OIXXIIXI, the other from the aquatic mammal GIIDII is used by us to season.

VEHICLES AND EQUIPMENT USED ON UMMO

INTERSTELLAR VEHICLES

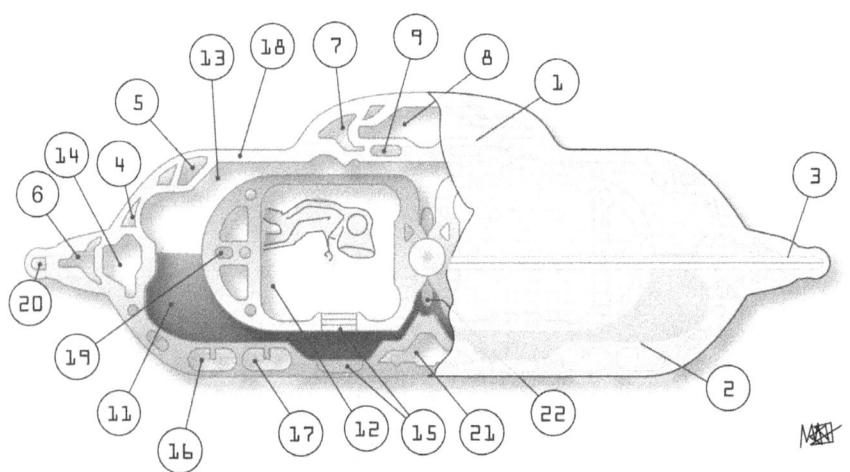

1 A protrusion, turret, or dome located in the upper hemisphere (it is transparent).

2 The central body of the spacecraft's superstructure

The entire superstructure of the spacecraft is protected by a finely perforated "ceramic" material that covers the outer armor. In addition, a 33-mm-thick safety layer surrounds the entire craft and separates the enclosure from the rest of space. All matter inside this enclosure can change its dimensional framework.

The structure's "ceramic" coating can alter its coefficients of elasticity and mechanical stiffness and features a vascular network containing conduits through which a liquefiable alloy and nanotools flow. These coefficients of elasticity can be adjusted at any moment based on multiple parameters dependent on the environment and the course of the flight. The hull must also withstand high temperatures caused by significant friction to which it may be subjected while passing through atmospheres of various chemical compositions and varying thermal conditions. It can also withstand continuous abrasion from cosmic dust and impacts from micrometeorites... It consists of a porous ceramic coating with a high melting point of 7,260.64 °C. Its thermal conductivity is very low. This allows it, for example, to be submerged in volcanic lava without any issues...

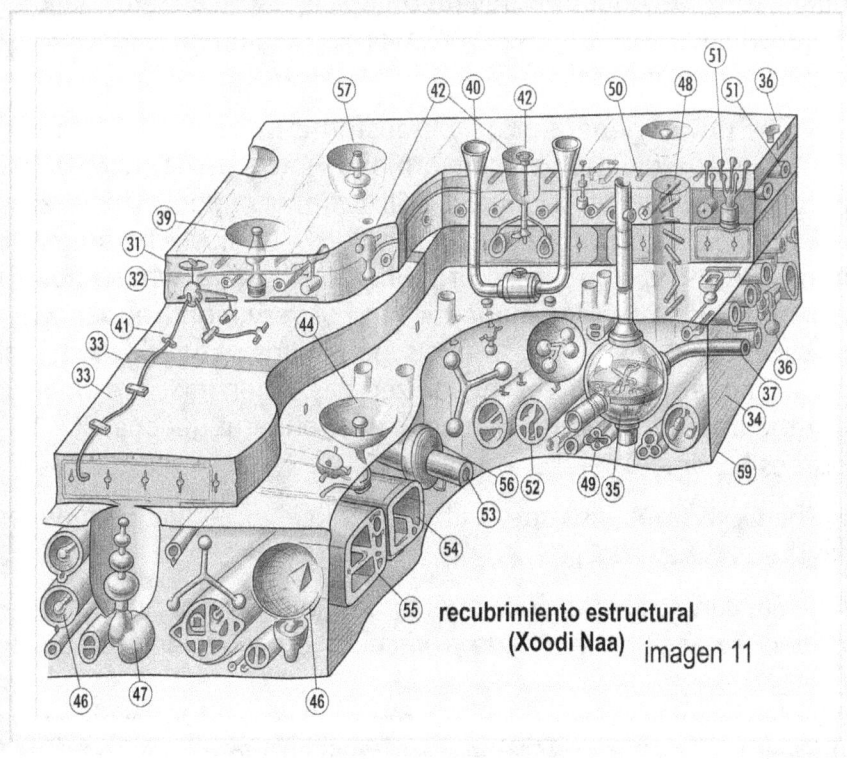

55 recubrimiento estructural (Xoodi Naa) imagen 11

3 - DUII: equatorial ring surrounding the UEWA.

4 - AAXOO XAIUU AYII: toroid magnetic field generator.

5 - NUUYAA: Toroidal tanks of oxygenated water and molten lithium.

6 - IDUUWII AYII: propulsion equipment located in an enclosure of annular shape embedded in the DUII.

7 - Power generator. Transforms the mass of lithium and bismuth energy, after its transformation into plasma.

8 - IBOZOO AIDAA: Control Centre of reversals IBOZOO UU.

"The equipment IDUUWII AYII (propulsion) is distributed, inside a toroid of revolution."

This term may be deciphered and transcribed as:

"The equipment for producing a 'frontier' force field (*) *within a toroid of revolution."

"The gravitational type high-frequencies are much less energetic than those of the electromagnetic type, although they are enhanced by an effect of gravitational 'self-resonance'. It's only for this reason that they are used for small applications and for domestic communications." (D 41-5)

"The power generator n ° 7, transforms a mass of lithium and bismuth in energy, after its transformation into plasma. (D69-1). The energy source is located in the ENNOI (turret or cupola). This energy generator also has a toroidal morphology. Its most characteristic element consists of a lattice of highly ionized gas whose flow is controlled by a complex magnetic field of very high frequency (in this case I use the word 'mesh' as a synonym for 'network' or 'spatial grid "). The temperature of the ionized gas, when it is in resonance with the frequency of the magnetic environment, reaches 7×10^5 degrees Kelvin." (D69-2)

The Oomomen also speak about a "nuclear" activator to generate the plasma.

"We can easily produce energy confining the antimatter suspended by antigravity inside a vacuum chamber and gradually

---

* "Cosmo-trampoline" effect on the cosmic layer XOODII, explained in PRESENCE—UFOs, Crop Circles and Exocivilizations

releasing its mass, making it crash into an equivalent mass of ordinary matter, then channeling the energy resulting from the merger process." (D1378)

THE CURRENT VEHICLE

The current vehicle is GOONIIOADOO UEWAA (flying vehicle). We wish to summarize the processes of traction currents on UMMO.

Except where traction in contact with the ground (as may be some applications in the movement of land, buildings hydraulic underground tillers, etc.) is necessary, the move people from one point to another of a communications network is done using two methods widely used on our planet.

. The first is a cataloged by us as YEDDO AYUU type (no network connection or organization). It is materialized by equipment that fit the human body, called OEMMIIUEWA and allows a person to move from one point to another in the air and low height (maximum height 30 ENMOO, some 56 meters). We use it when we have to travel relatively short distances but almost never for short trips in which case we use as you, feet. Travels in these conditions (we refer to personal equipment OEMIIUEWA) are not controlled by XANMOO AYUBAA (computers network). Instead, the individual who uses it must control its direction continuously as you usually do with your cars, with the only difference that we do not use manual controls (such as the steering wheel, pedals, buttons, etc.) but merely codified acoustic orders

. But the vehicle by antonomasia (*) as used on our planet is probably the GOONIIOADOO UEWA (picture 16). Its shape may remind you of some futuristic devices designed to travel or that of some modern cars with aerodynamic profile.

---

* Appears to replace the name with a quality of the thing or being designated. For example "automobile" used as a common name for his automobile properties—TN

picture 16—GOONIIOADOO UEWA

The process of traction used is based on an ancient principle for us, but that has not been changed essentially for the good reason that his system, which has nothing to do with the one we use for our sidereal trips in the OAWOOLEA UEWA (cosmonefs), is very effective for traveling long distances in the atmosphere.

The traction system consists of a BUUTZ (motor) GOONNIAOADOO (special state of matter that is not solid, liquid or gas), and whose only fuel is the Xenon tetrafluoride.

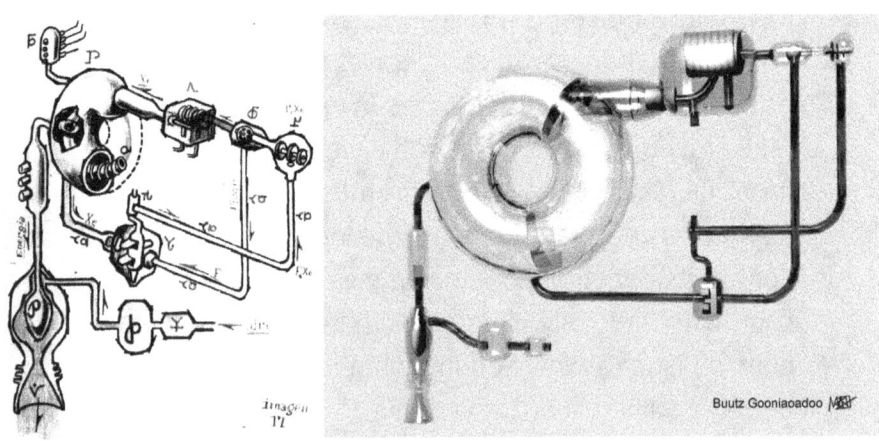

These vehicles are moving at very low altitude, always avoiding accidents of geography and is now always at the same height in accordance with the variations of natural and artificial (at 0.3 ENMOO or 0:56 m above the ground) so that even accident—hazard which we reduced the probability to 0.00007—travelers do not suffer significant injuries.

Its remote sensing is achieved, controlled by the XANMOO AYUBAA (worldwide network of computers) simultaneously with detection equipment that controls the vehicle at all times not only the meteorological parameters and the presence of static obstacles, but also the likely direction of other vehicles traveling in the immediate. It also prevents the presence of XAABII (Houses) "buried" that may emerge quickly by colliding lamentable.

VEHICLES OF THE PAST

Have disappeared from our orography, the old paths or tracks on which circulated in the historic time, these old NOIA UEWA (picture 18) which traveled through hinged feet (anthropomorphism of the technique prevented the routine use of the wheel as a means of movement) on pavement or trails that are different from the roads since they were built by stabilizing the field by additions of clay composition significantly, silicates and synthetic alumina for the wearing instead of being constituted as the motorways of the Earth using conglomerate of aggregates and asphalt materials.

picture 18—NOIA UEWA

EXAMPLE OF TECHNICAL FLYING VEHICLE

The XOOIMAA UYII UULUEWAA are flying equipment for geological work.

They have the shape of an ellipsoid of revolution. Equipment used on our planet is far greater. The device has a multiple frequency generator in the range of 5x1020 cycles per second, in addition to some temperature sensors, magnetic field strength and gravitational. The device is controlled by a frequency beam of 6.77 x 1020 cycles/second.

It settles to the ground and begins to build around a gaseous layer of GOONNIAOADOO (you call it PLASMA). The high temperature of the outer skins of the plasma layer magnetically controlled, capable of spraying the silica, make sinking the device within the different layers or strata of the ground like a coin hot sinks in an ice bar. (See picture 2)

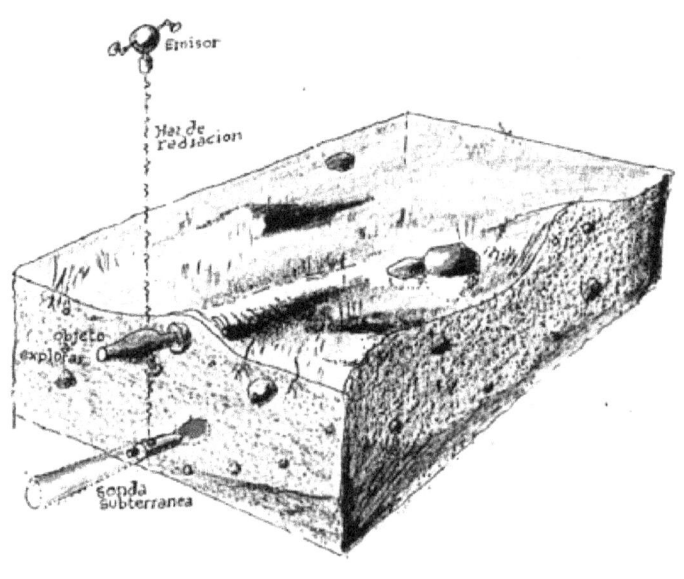

picture 2

Once buried at depths of about 50 to 100 meters, it is controlled to be able to move horizontally (picture 2A), occupying a series of points corresponding to an ideal network, from which it emits a beam tronconical of waves (frequency cited) that passes through different layers. Simultaneously a UULUEWA located in the air at a height of 20 or 30 meters picks in all other points in the network the pulses emitted by the device that sends its from his great depth.

The system has a certain resemblance to the radios-metallographic you use to examine the metal with the Rœntgen rays. The difference is that instead of using a giant screen to save the X-ray from the field, an explorer air sweeps all points of the ideal picture by recording and integrating them in a crystal titanium memory that will give us at the end the three-dimensional view of the Earth's crust within explored with all its internal accident contrast achieved with their opacity to the passage of these rays.

With miniature equipment like those used by our brothers in India, one can discover and veins of minerals, water currents or hydrocarbons, buried rocks, tunnels and pipes buried ruins of prehistoric cities and different objects larger than 5 cm (with miniature equipment such as our brothers used it in India).

Morphology of the Oomomen

The average size of women and men from UMMO are respectively 195.0 and 208.2 cm. The standard deviation around these mean values is 4.5 cm for women and 4.8 cm for men. The expeditionary detached on OYAGAA are normally selected from men of smaller size to one ENMOO (about 187.4 cm) and women with a size less than 179.4 cm.

Given the small number of OEMMII which has both a sound organ functional and appropriate size it is difficult to satisfy these criteria. So, AYIOA 1 son of ADAA 67, which I depend, is more than two meters high, other reasons prevailed for his participation in this shipping. It is difficult for him to mix with people of the countries of southern Europe but it is less problematic in areas at north where his size is not exceptional. Our body size slightly exceeds yours with a body mass index mean varying between 25 and 30.

An Oomomen's couple—Album "Like 2 OEMMII" of Ummo Music
www.ummomusic.com

We can endure, without protective clothing and without muscle activity; temperatures that can be lowered to 2 °C if we are sheltered from wind and rain. The temperature inside our homes

is generally controlled between 8 °C and 14 °C. We support the heat up about 28 °C but cannot effectively remove the internal temperature excess only on limited terms and if we leave exposed to outdoors much of our skin, which is impossible on your planet. Our skin takes on a slightly tanned color as if we were covered with suntan oil.

This excess of exudation results inevitably an excessive pherormonal release that excites your animals, especially dogs and flying insects. We are reducing our activity on OYAGAA during the summer months when we focus on operations contacts and scientific analysis within an adequately conditioning.

THE POLITICAL ORGANIZATION ON UMMO

STRUCTURE OF OUR AYUYISAA

The OEMII (human being) of our UMMO planet enjoys a greater freedom than on any Nation on the planet Earth. On being

UUGEEYIE (child), it is considered as the beneficiary of the basic rights that are EQUAL IN PRACTICE FOR EVERYONE WITHOUT EXCEPTION. It is only when the individual violates CONSCIOUSLY (and we have effective ways to differentiate compulsive acts of those who are enabled by the free will) HIS BASIC OBLIGATIONS or mutual respect for his brothers and he deserves a penalty for his transgression serving the community obediently.

Our Laws are especially severe in those cases where the basic rules of our SOCIAL NETWORK are violated. All the brothers of our planet are governed by three Supreme Government organizations whose powers reach several levels and areas.

. The UMMOAELEWEE Group usually composed by four members of the both sex

. The UMMOALEWEEANNI Group formed by 116 GEE and YIE as maximum figure.

. The UMMOOEMII. Group .composed of all OEMII (humans) in full possession of their faculties psychosomatic. It is difficult to summarize the degree of influence and competence of each group of the AYUYISAA (state). We obtained up to now a serene balance of social networks, very stable—even if it can be considered it still can be improved—by assigning to each assignment and essential functions.

The UMMOEMII has the ability to create the laws of our planet. Periodically, each group of 12 people self-selected by explicit vote freely choose a representative (AOUIAOEMII). In turn, 1,728 of them elected by secret vote one OUIAOEMII who has the right to propose the creation or modification of the UAA (laws) and to vote in the UAAYUBAA.

This is an UAAYUBAA located in the region of OAROO AAXAA at 3.62 KOAE from the largest colony of that name, situated in a vast forest of NAANAA and OBUANAA (unknown large trees on Earth, with cardioids leaves). All OUIAOEMMII live nearby with their family in XAABI (House Tours) which are distinguished by their yellow color. In the center of a valley is the metal structure of a large hemispherical building.

Actually you can compare our OUIAOEMII your parliamentary or representatives of the Earth, and such a comparison is plausible because they are also legal representatives freely chosen by the AYUYISAA (human group). But their functions and working methods are different from those of their colleagues on Earth. The UAAYUBAA bears no resemblance to traditional terrestrial parliamentary halls. Rather, it is a veritable laboratory study, with connections of terminal equipment or XANMOO AYUBAA (computer network that spans all UMMO). The OUIAOEMII are real scientists who work all day for the study.

Every day they get there, from any part of UMMO, millions of acts of sociological type on the conduct of all the brothers of UMMO. These are statistical facts collected by the XANMO AYUBAA, but do not believe that the UMMO brothers feel slaves day and night from a species of mechanical and ruthless spy who saves all their reactions, disrupting their freedom. This recording exists, yes, but what we realize does not interfere because the Xanmoo Ayubaa does not record the names of OEMII because those analyses only count the overall statistical figures (we do not realize studies by samples).

Anonymity is therefore perfect when those to whom it refers. Such facts are carefully recorded and used to assess the extent to which a UAA (ACT) continues to be fair (remember that humanity is evolving), and therefore must be changed or canceled. But these complex facilities provide only quantitative results. The final laws and important decisions must be adopted by the specialists. They are among many possible solutions those which have a higher level of statistical viability. That's when, well informed, the OUIAOEMII seriously discussing and vote the decision to adopt.

Life on Ummo—artist representation—Davy H.

For example:

There used to be a UAA (NORM) in XEE (years) 317 of our time, whereby the UUYIE (girls undergoing puberty) who had not attained 17 years of Earth age who were affected by their OAGOOU (menstrual period) were not required to carry out their normal school work for one period before and after EIWOO ⟨X [Ovulation: Critical phase evaluated from -1,270 to +2,380 UIW (*)], because the studies previously shown the risks that may lead to future to profound alterations subconscious psychic type.

But the 317/26 604 🦋 on adopta la nouvelle UAA en diminuant ces intervalles : – 1 106 UIW à + 1 875 UIW (1 UIW = 3 092 minutes).

was adopted the new UAA by decreasing these intervals: -1,106 to +1,875 UIW (1 UIW = 3,092 minutes).

The facts reported psychological and physiological type corresponding to all UUYIE (girls) in UMMO, showed that the probability that happens such disorders had become smaller. And that UAA this continued to be transformed so far. Today we apply these standards with different time intervals for each child based on his own personality and circumstances.

The UAA (laws) from this organ (UMMOEMII) must be followed by ALL of our UMMO brothers without exception, even by the four paramount UMMOAELEWEE chiefs. Never a OEMII can not allege ignorance of the law, because when he has doubts in some concrete cases, it has at its disposal a connection with the XANMOO AYUBAA (Computer network) that provides phonetically and graphically all the information on their Civil rights and obligations to the State.

At 13.6 years (terrestrials), each UMMO child, without exception, undergoes an analysis (the third of his life) much more comprehensive than previous ones, in a special "psychological studies" center. It evaluates all his faculties, mental, intellectual, but also on the emotions and the neural structure.

This diagnostic is not achieved with standard psychometric methods of Earth specialists. Excepting the anamnesis provided for each exploration, the study is performed by evaluating a series of biological functions carried out by the electrostatic and magnetic

---

* TN: about 2 days before and 4 days after

fields created by the nervous system of the subject (this method has a distant resemblance to the electro-encephalograms well known by terrestrials' neurologists).

In this way we select a total of 120 children whose UMMOGAIA DO DA (psychosomatic identity formula) presents the features of a high capacity (you will say, individuals who have the biggest or the higher IQ or more able). These children, regardless of gender, are intended for ONAWO WUA (TEACHER center) where they receive special training adapted to the transcendental functions they need to develop. There is no problem of maladjustment to such studies because their selection was made specifically with respect to such skills. In addition to the Graduate School of UAA, sociology, psycho-biology, etc.

It instills a deep spirit of particular service to their UMMO bro-thers. They will come to have great authority and they will be res-pected by all OEMII on UMMO. At the same time they should see themselves as being in the service of everybody. They know that a simple observation on their part causes predictable reactions of submission of their environment, but it is prohibited to humi-liate just for the pleasure of being obeyed. In usual circumstances of their lives, they must accept instead the most menial tasks and the humblest. If they do not have enough strength to adapt to this situation, they confess modestly and they are respectfully excluded from the group. Once completed this training, the duration is not the same for all group members, we realize a new selection of four members who by their exceptional ability will compose the UMMOAELEWEE as we will explain further. Theoretically, there are 116 members who will form the UMMOAELEWEEANNI immediately.

The UMMOAELEWEEANNI COUNCIL has two prerogatives.

One: OMMOI DOOXOO (you could translate by JUDICIAL) and the other, to ensure that the balance of forces is maintained in the General Council of UMMO (the four members of UMMOAELEWEE).

They ensure that the UMMO UAA (laws) are met, and they inter-pret the transgressions of UAA (laws) by applying the sanctions. Their judicial function has, however, peculiarities different from the judicial-legal organizations as on the Earth Planet.

First to rule in accordance with the UAA, they reflect not only the testimony of the brother under their jurisdiction, but also of facts supplied by the XANMOO AYUBAA. Do not interpret this as a judgment of XANMOO AYUBAA. It would be so monstrous to be slaves of the own OEMII mechanical creations.

The assistance of XANMOO AYUBAA, though far more objective and scientific with a high degree of accuracy, can be compared with the occasional help of a camera or tape recorder in front of terrestrial Courts. Such methods are infinitely more impartial and accurate than the testimony of fallible human beings who can be polarized by hatred or emotional imbalance.

There is the POWER OF THE LAW as an expression of the will of all people over even the largest organ of the Nation (the UMMOAELEWEE).

Contrary to what is observed in Totalitarian States, the legislative, judicial and executive branches are not subject to the same commanding group. In our case they have full autonomy.

And when later we make a short presentation of our basic rights, you will see that and the FUNDAMENTAL LIBERTIES and RIGHTS ARE INSURED for all OEMII on UMMO.

For this reason in this framework we will present two summaries:

. First, declaration of human rights accepted by the UN and finally ratified, after centuries of intolerance by the Catholic Church of Rome.

. Second, our series of protective UAA's OEMII that, globally considered, can be compared to previous terrestrials' principles.

You will observe that apart from certain peculiarities, the analogy is very significant.

SUMMARY OF THE DECLARATION OF HUMAN RIGHTS (Universal Declaration of the UN: European Convention: Pacem in Terris: Vatican 2)	SUMMARY OF THE SERIES OF UAA PROMULGATED FOR OEMII PROTECTION ON THE UMMO PLANET
SUMMARY OF THE DECLARATION OF HUMAN RIGHTS	--------------
(Universal Declaration of the UN: European Convention: Pacem in Terris: Vatican 2)	The OEMMII is free to join any Thought, Science and Technology criteria.
-------------	The physiological structure of OEMII is inviolable
RIGHT TO LIFE AND INTEGRITY OF MAN.	The OEMII can freely integrate all INNAYUYISAA (Small SOCIAL NETWORK or GROUP)
RIGHT TO FREEDOM OF BELIEF, THE RIGHT TO PUBLIC WORSHIP EVENTS IN EQUAL TERMS.	
RIGHT TO FREEDOM OF ASSEMBLY AND ASSOCIATION. RIGHT TO POLITICAL PARTIES AND FREE TRADE.	The OEMII is free to adopt any active function considering that its effectiveness will be reduced if it does not accept professional guidance offered by the UMMOAELEWEE.
RIGHT TO FREEDOM OF MOVEMENT AND FREEDOM OF RESIDENCE. INVIOLABILITY OF CORRESPONDENCE.	The OEMII may require AYUYISAA (SOCIAL NETWORK or SOCIETY) to provide in any case: XAABI (residential), UAMII (ADEQUATE FOOD), training in ONAWO WUA (teaching), AARGIAGOO (PREVENTION AND MEDICAL ASSISTANCE), WOAII OO (RELIGIOUS SUPPORT) XANMOO AYUBAA (incidentally using computers).
RIGHT TO OBTAIN A STANDARD OF LIVING ECONOMIC-SOCIAL SUFFICIENT. RIGHT TO STRIKE. RIGHT TO WORK.	
RIGHT TO LEGAL INSTITUTIONS OF A DEMOCRATIC STATE. RIGHT TO PARTICIPATE IN FREE ELECTIONS.	

RIGHT TO EQUALITY IN FRONT OF THE LAW, REGARDLESS OF PRIVILEGE CLASS OR GROUP.

RIGHT NOT TO BE HELD ARBITRARILY AND TO BENEFIT THE PROPER LEGAL PROTECTION.

We have synthesized some rights that distinguish one Totalitarian State from another regarded as legally RULE OF LAW, following the Earth convention. The lawyers in Law Policy consider these notes as objective evidence to describe the immorality and illegality of a terrestrial social structure or on the contrary, accept the degree of legality.

The OEMII can communicate as he wants with another OEMII by telepathic call or by any other physical means.

The OEMII is free to access any information within the AYUYISAA

The OEMII is free to require from a hierarchical chief the reason for his conduct and rationally discuss his orders.

All OEMII is free to integrate without any constraint into a INNAYUYISAA (small group) whose members consciously renounce to all the privileges of the OEMII expressed previously, accepting as ultimate reason for the good of AYUYISAA (all UMMO other brothers), rigorous discipline and total dedication of its physiological structure above.

All OEMII, as the owner of an aliquot part of the planet, may require from government organizations, the degree of enough freedom to enjoy it.

Afinh   *Itinen.*   *bien*

El túnel y el Jardín ya existían desde su generacion. Cuando el viajero piensa que lo que en ese momento estáenfocando con su linterna, (PRESENTE) se acaba de generar en ese instante, o que lo ya visto ha dejado de EXISTIR o que aquellos muebles que verá aún no existen ¿ No es víctima de una cándida ilusión....?

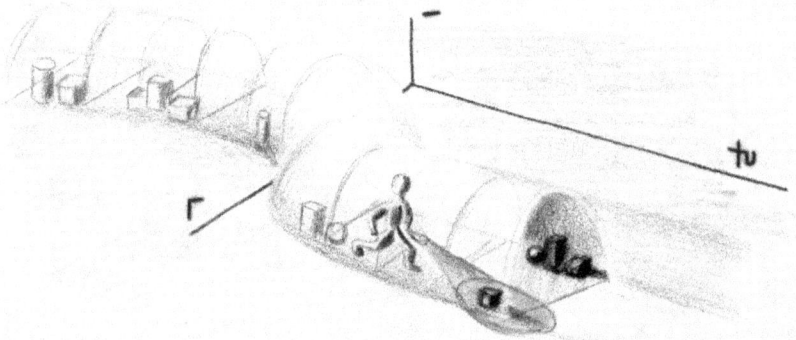

337 LIBERTAD EN EL ALBEDRIO DEL HOMBRE HUMANO.

Cuando este concepto del OEBUMAEOEEMIII empezó a esbozarse en   1

· · · · · · · · · · · · · · · · · · · como verdad cietífica) surgió una

T.6   24

Ustedes han estudiado que todo nucleo atómico está rodeado de una nube de electrones, situados a distintos niveles de energía. (Vease el ingenuo gráfico que lo simboliza) IMAGEN A

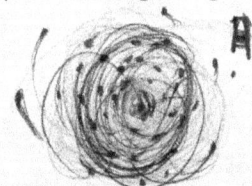 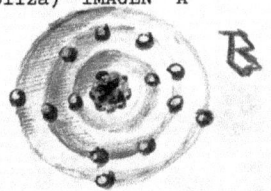

Dentro de cada nivel la situacion de cada electron es una funcion de probabilidad, es decir del AZAR (Recuerden el principio de INDETERMINACION o INCERTIDUMBRE.

Cuando los libros elementales de ustedes describen a un átomo simbolizan las distintas "CAPAS" o NIVELES DE ENERGIA como en la IMAGEN B.

En UMMO a los niños le ponemos el siguiente SIMIL sencillo con fines didácticos, que expresa mejor este concepto a los no versados aun en Física Nuclear.

(a pesar de estas revelaciones) al nenos juzganos que pueden servirles a ustedes para adenas de incrementar su cultura, corregir en parte sus nismas estructuras personales.

337 UNION DE TODOS LOS HOMBRES A TRAVES DEL BUUAWE BIAEI

Veamos en un primer esquema elemental como estan interconexio nados los quatro factores integrantes de la personalidad de dos se res humanos:

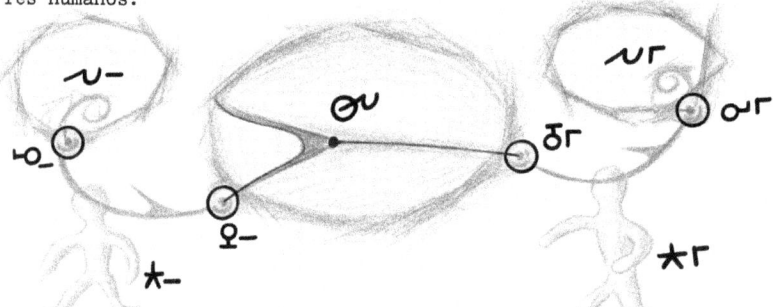

Un OEMII (CUERPO) ✶– Correspondiente al Hombre Nº 1 Desea estable cer comunicacion telepática con el OEMII NUM (CUERPO HUMANO "2") ✶ᴦ Analicemos superficialmente el proceso:

que desconociamos.

Nuestra BUUAWEAA posee conciencia Mas no identifiquemos esta CONCIENCIA con la CONSCIENCIA del OEMII. Explicaremos esto, describ do lo que ocurre en nuestro cuerpo cuando miramos por ejemplo la imagen de un triángulo. Ilustraremos el proceso por medio de grose- ras imágenes trazadas con el auxilio de lápices terrestres con mina coloreadas.

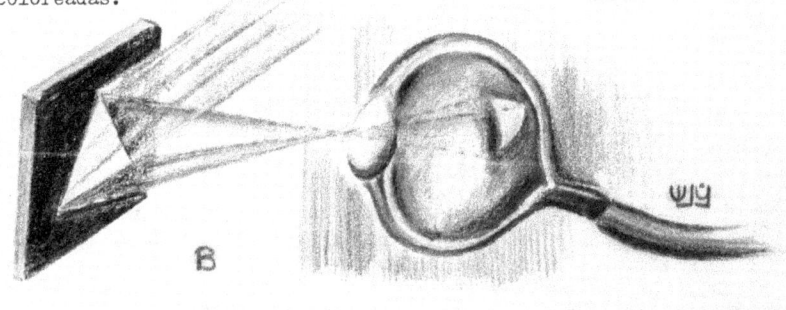

B

Sobre una lámina B en la que hay dibujado un triángulo blanco, sol fondo negro, incide un haz (rojo) de IBOAAIA OU (Fragmentos cuanti: cados de energía, con característica ondulatorio- corpuscular) que

pues imaginar, la información codificada que será capaz de acumular.
Ninguna etra base MACROFIFICA de MEMORIA puede comparársela.

Los bloques de TITANIO que utilizamos han de presentar una estructu-
ra cristalina perfecta y un grado de pureza química de rendimiento 100%.
Bastaría la inclusión de unos átomos de impureza (hierro, molibdeno, si
licio ...) para hacer inutilizable ese bloque.

Ustedes pueden preguntares ¿como es posible el acceso, a uno por uno de
esos átomos del bloque, para codificarlos excitándolos ó extraer la información (decodificación)acumulada?.

Un esquema ó dibujo elemental aclarará ideas.

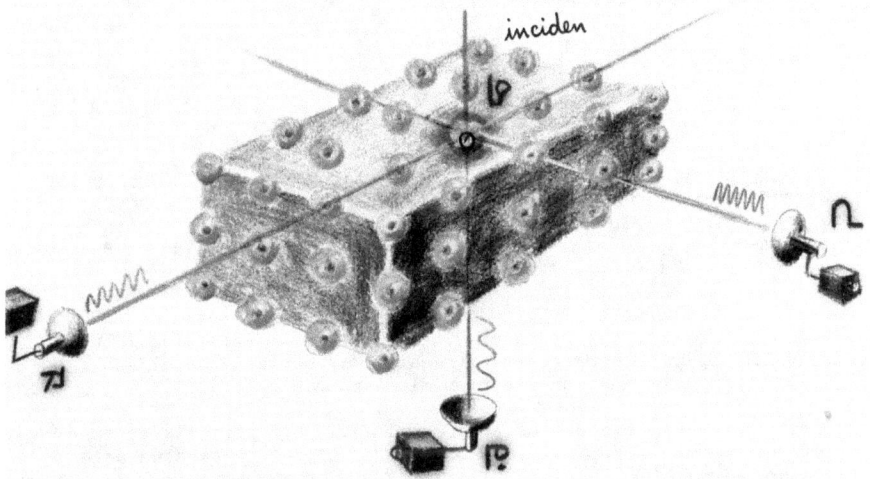

Sobre un bloque de TITANIO inciden tres haces (simbolizados en el dibujo con los colores carmín, azul y verde) de socción infinitesimal y fre
cuencia elevadísima, (capaces por tanto de atravesar el bloque sin afec
tar los núcleos de sus átomos, (pero sí las cortezas electrónicas respec
tivas) se utilizan por ejemplo frequencias del orden de (8'35.10 cicl s
por seg) y distintas para cada haz. (IO|7J|ᒋ) son los generadores de
frecuencia.

Estas elevadas frecuencias caen fuera del espectro característico del TI
TANIO por lo que esos haces independiente considerados no son capaces de
excitar uno a uno, sus electrones corticales.

Más no ocurre así cuando los tres reyos inciden simultáneamente sobre un ATOMO –
específico (el IᎬ del dibujo). Entonces la superposición ó mezcla de
las tres frecuencias provoca un efecto de antiguo conocido por ustedes,
llamado BATIDO ó HETERODINAJE que dá como resultado una frecuencia mucho
más baja y que coincide con cualquiera de las rayas espectrales del TITA
NIO.

El átomo es así excitado y como los tres haces ortogonales pueden despla
zareo en al espacto con gran precisión. Localizan uno a uno todos los at

T.5   27

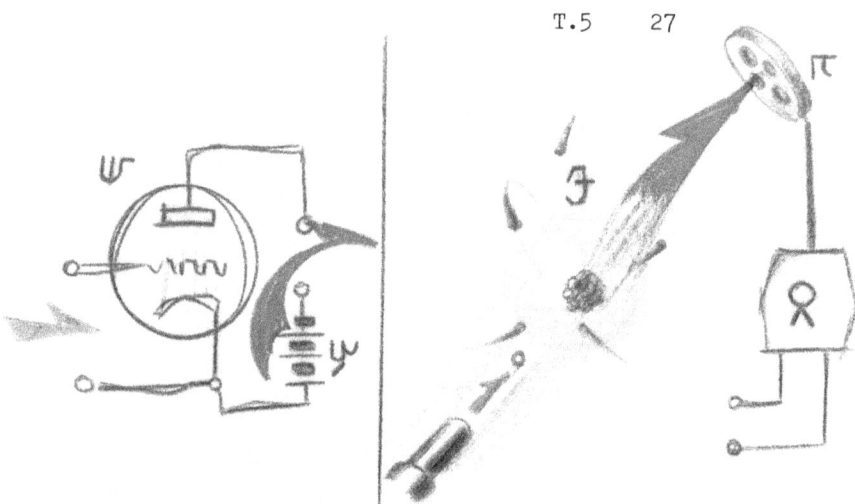

Observen que una energía de entrada (Flecha verde) puede controlar una gran energía (Flecha roja), pero !solo controlar!. No genera - energía; controla aquella energía eléctrica proveniente de la Bate_ ría (𝝭).

Por el contrario, en el ODU GOAA, una debilísima energía (NEUTRON) (FLECHA VERDE) provoca una escisión nuclear en un solo átomo 𝔍 cuya fisión libera una enorme energía (Fecha roja) captada por el AASNEII (𝔽) y transformada, de calor en electricidad en (𝝦).

En principio, este proceso es análogo al utilizado por ustedes en - reactores nucleares ó pilas atomicas, pero controlado para un solo átomo en nuestro ODU GOAA.

En los computadores digitales de Tierra, unos equipos denominados uni dades aritméticas, realizan a gran velocidad operaciones elementales (sumas, restas ...) empleando módulos transistorizados.

UMMO utiliza IVOAWE BOO, basados en reacciones químico-nucleares a es cala microfísica en vez de transistores.

Para ello empleamos unos pocos centanares de estas reacciones básicas, elegidas específicamente, de modo que los dígitos utilizados seon ex-presados en el sistema de base 12.

Por ejemplo: la codificación de esta suma: y su correspondiente verifi cación.

$$12 + 1 = 13$$

Se realiza por medio de esta reacción. (En la que intervienen, no

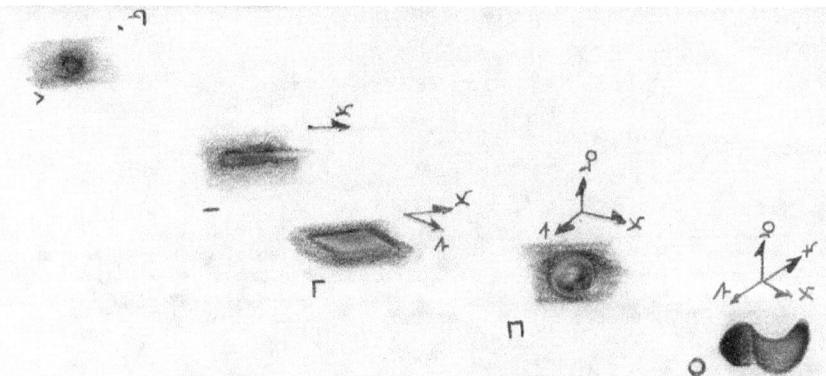

Pueden ustedes imaginaros que nuestro BICOSMOS primitivo, se aseme
jaba mas a una poqueña esfera vacia. Un universo diminuto sin Gala
xias, sin gases intergalacticos... solo espacño existiendo en el
tiempo.

WOA curva una y otra vez este espacio. Cada curvatura nueva su-
pone una dimension y por último lo "arruga" Observese que estamos
empleando un simil, un simbolo, pues solo matematicamente podrfamos
expresar lo anterior con genuina corrección Por ejemplo: la expresion "arrugar el espacio" parece infantiloide pero en muy dãdactica

Con una mueva imagen intentaremos hacernos comprender

Si un espacio tridimensional lo curvamos, arrugamos, o hacemos
una especio de hoyo como ven en la figura, a traves de una cuarta
dimension, esta curvatura en la que nuestros organos sensoriales
interpretarán como una MASA (Una piedra, un Planeta, una Galaxia)

Pues bien. WOA extorsiona aquel,microbicosmos generando así la ma
Nada menos que easi toda la masa actual de muestros dos Universos
gemelos, concentrada en un espacio reducidísimo. Algo así como toda
el agua carbonatada de UMMO encerrada en mi puño. Materia y antimateria como untedes la lluman, superconcentrada.

Se produce entoncos una doble EXPLOSION-IMPLOSION. Por la IMPLO-
SION materia y antimateria, es decir, atomos positivos y atomos
negativos son atraidos violentamente unos contra otros sin encontra
se jamas. Son dos conjuntos, dos Universos, WAAM yUWAAM que jamas
podran encontrarse or que no les separan relaciones de espacio.

De modo que cuando decimos que se atraen, el verbo "atraer" debe
interpretarse como"se interinfluencian"

Por otra parte indicamos que se realizó una explosion. En efecto:
La immensa masa de cada Cosmos se fragmenta en particulas , y estga

vivira un hombre a lo largo de su VIDA

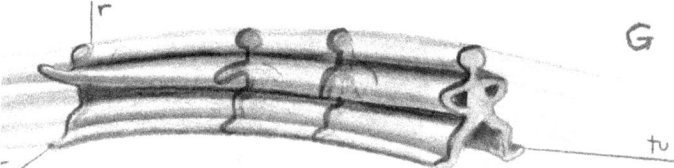

Cada situación lleva asociada una FECHA (IMAGEN F) Pues bien: El espacio y el tiempo estan asociados tan estrechamente, que si unimos en
una misma expresion gráfica, en una sola imagen, todas esas situaciones
o sucesos que vive el hombre a lo largo de su vida, obtendremos un ex-
traño SER de cuatro dimensiones (volumen-tiempo) parecido a un enorme

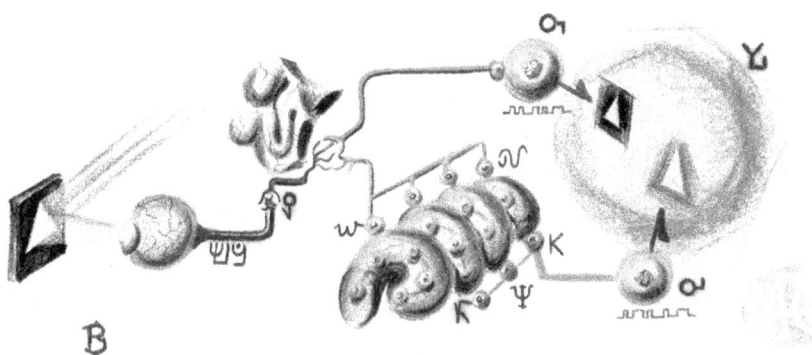

OEBUMAEI (Especie de "churro" o "embutido" mantecoso y dulce, cuya
sección representa un muñeco al partirlo en lonchas, y que es muy

          región de AADAAADA (UMMO).

          hacemos un resumen del informe remitido a dos especialis-
tas japoneses.
    Las neuronas de esta última via aferente, sinapsan a su vez cada
una a dos fibras nerviosas. YAAIODEE (verde) y YAAIOSUU BIIO (azul)
 Dicho de un modo sencillo: Cada impulso nervioso que representa
un punto del TRIANGULO, es transmitido a la vez por dos vias —como
si una de sus telefonistas enviase un mismo mensaje a dos corresponsa-
les.

Los impulsos de la via YAAIODEE (verde) operan sobre el organo de la
MEMORIA. cuyo funcionamiento parecen ustedes ignorar. La imagen simboliza al BIEYAEYUEO DOON(dibujada de color ocre) Se trata de una

Un dispositivo (**ℛIʃ**) fija opticamente la imagen ondulada por medio de carbono pulvurulento con un aditivo aglomerante.

(**τόⱢ**) es un calefactor a METANO que fundía la mezcla estabilizando definitivamente la función acustica grabada.

Con este rudimentario sistema que guarda un lejano parecido con las actua—les técnicas de grabación foto-optica en "film" Tierra de tipo cinemtográ fico nos ha sido legado el conjunto cultural de la época.

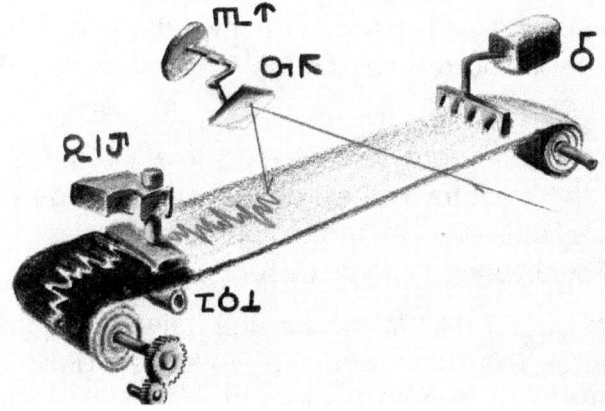

El agua fría por unas tuberias (**ʊⴲⵢ**) ubicadas a lo largo del eje y era calentada por las radiaciones infrarrojas de IUMMA hasta su vap ción. De ese modo podía aprovecharse aunque con rendimiento inferior por ciento,para su conversión en energía mecánica.(Ved DIBUJO)

NOTA CUARTA

Existian ya rudimentarias técnicas de grabación sonora.

67

ORIGINE ASTRONOMIQUE DE NOTRE COMPUT TEMPOREL

Notre manière d'évaluer les grandes périodes est différente de la vôtre et celle-ci s'est maintenue tout au long de notre histoire, ayant son origine dans une très ancienne mesure astronomique. Nous définissons parfois, à tort, le XEE (« année » d'OUMMO) comme le tiers de la période de révolution de notre OYAA OUMMO autour de notre soleil IOUMMA. La valeur du XEE est de 77,3 jours terrestres.

IOUMMA est une étoile dont de masse égale à 1,48 x 1030 kilogrammes. Son spectre lumineux est décalé vers le rouge en comparaison à votre Soleil avec des indices photométriques U-B et B-V égaux à 1,15 selon votre système de référence.

OUMMO gravite autour d'IOUMMA sur une trajectoire quasi circulaire d'excentricité 0,0078. La distance moyenne OUMMO-IOUMMA est de 9,96 x 1010 mètres. Un autre OYAA de taille importante, NAWEE, gravite autour d'IOUMMA sur une trajectoire elliptique d'excentricité 0,026, à une distance moyenne de 5,97 x 1010 mètres.

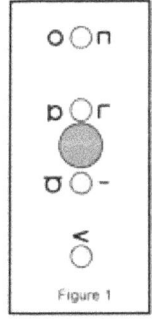

Figure 1

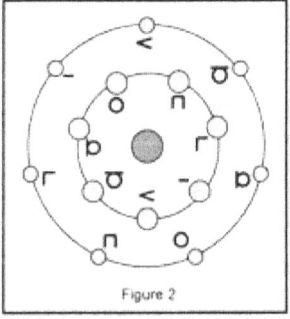

Figure 2

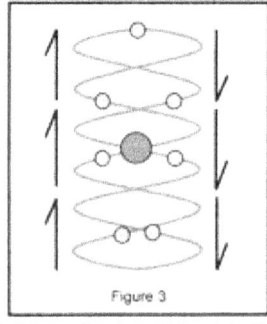

Figure 3

Les schémas ci-dessus, à visée didactique uniquement, sont volontairement ramenés à des configurations idéales et simplifiées

Les anciens astronomes utilisaient comme références les deux astres IOUMMA et NAWEE, ce dernier étant bien visible conjointement à IOUMMA au crépuscule et à l'aube. Les astronomes observaient les passages en conjonction supérieure de NAWEE en alignement avec IOUMMA, ce qui se produit en moyenne tous les 2 571 XEE, soit 0,866 du temps de révolution sidérale d'OUMMO. Lors de chaque conjonction, ils notaient une variation verticale importante de la positon de NAWEE par rapport au disque d'IOUMMA (voir figure 1), l'oscillation se réinitialisant chaque XEEOUMMO — dix-huit XEE — lorsque NAWEE avait accompli treize révolutions complètes et OUMMO six révolutions (figure 2). Les cosmologues, ignorant que le plan de l'écliptique d'OUMMO possédait une orientation distincte de celle de NAWEE, en conclurent que la trajectoire d'OUMMO était bi hélicoïdale et qu'OUMMO décrivait trois translations descendantes et trois autres montantes, sur la surface fictive d'un cylindre, pour compléter un XEEOUMMO (figure 3).

Le XEE peut donc se définir comme le tiers de la période de révolution d'OUMMO. Il existe cependant une erreur minime dans cette définition car les anciens cosmologues d'OUMMO vouaient majoritairement un culte au nombre π. Ils voulurent absolument faire correspondre la valeur du XEE à la fraction $\frac{2}{2+\pi}$ du temps de révolution synodique de NAWEE en acceptant une dérive de six OUIW (18,55 minutes) chaque XEE. La dérive cumulée atteint ainsi un jour d'OUMMO tous les 100 XEE et un XEE tous les 6 000 XEE. Les adeptes du culte de NAWEE, indignés de cette offense faite à la rigueur objective, promulguèrent aussitôt un édit eschatologique annonçant la fin des temps au bout d'un cycle de 6 000 XEE si l'erreur n'était pas rectifiée. Le monarque de l'époque trancha en faveur de la majorité. L'erreur fut ignorée et se perpétue encore de nos jours. Quelques projets furent proposés par le passé pour influer sur l'orbite d'OUMMO afin d'annuler la dérive, mais les différents OUMMOAELEWEE auxquels ils furent soumis les refusèrent systématiquement.

Hormis ces considérations concernant les XEE, une autre dérive existe dans la définition du XI (jour d'OUMMO) qui correspond en pratique à 600 OUIW. La valeur réelle atteint toutefois 6 000 117 OUIW. Cette approximation sur la valeur du XI implique logiquement un décalage progressif de l'horaire du lever d'IOUMMA en un point fixe donné de l'équateur d'OUMMO. Lorsque l'OUIW fut introduit durant l'essor scientifique que connut notre peuple au cours de sa seconde ère historique, la valeur officielle du XI fut redéfinie et une tentative de réajustement calendaire eut lieu pour lui conférer une meilleure corrélation avec la réalité astronomique Un OUIW fut ajouté au comput temporel chaque 84 XI, à l'exception d'une fois tous les 84 XEE. Cette méthode fut rapidement abandonnée car jugée inutilement complexe. Nous n'avons pas, sur OUMMO, l'impératif d'un calcul calendaire précis pour savoir à quel moment effectuer l'ensemencement de nos cultures en fonction des conditions climatiques à venir. Les conditions climatiques sont sur OUMMO, entièrement dépendantes de la latitude, et en aucune façon du positionnement physique d'OUMMO sur son chemin orbital.

Nous joignons à cette lettre la traduction française de la réponse que nous avons donnée récemment à l'un de vos jeunes frères de Freie Hansestadt Bremen qui s'interrogeait sur l'existence d'années bissextiles sur OUMMO. La présente lettre représente un complément à cette réponse dont elle reprend quelques passages.

*

THE MYSTERIES OF THE UMMO FILE DISCLOSED

There is a rumor that the Oomomen have put false information in their documents to mislead readers. These documents were written over 50 years and deal with multiple topics ranging from logic to astronomy and cosmology, medicine, and the presentation of new physical paradigms. However, it is interesting to note that in reality the documents contain very few errors and text mistakes, and they rarely have spelling mistakes, no matter what language they are written in.

> D32 18/03/1966 ESP: « But remember that a mathematical error, a conversion of physical units misinterpreted, can discredit us from the experts. In this case we would have no credit with these people. The skepticism of your readers would be totally justified because no one is obliged to accept testimony that is based on contradictions or statements devoid of any objective scientific aspect. On this aspect we are very understanding. We cannot demand to be believed without providing convincing evidence."

THE STAR IUMMA FINALLY DISCOVERED

A ploy often used by the "debunkers" of the Ummo file is to present supposed false information that will prove that the whole Ummo file is a hoax. This rhetorical nonsense is applied to the location of the star "Youma" (IUMMA), where the Oomomen say that they come from. In the documents they talk about the star Wolf 424.

In fact, the Oomomen indicate in the document referenced D21 "We're not sure whether it's the same star" ...IUMMA localization is difficult due to a cloud of cosmic dust which reduces the brightness of the star making it seem further away. The brightness of the star is expressed by a measure of visual magnitude. The higher this magnitude, the lower the brightness of the sun. If a visual magnitude is very high, it might seem like the star is a long distance away.

« You evaluate the absolute visual magnitude in a conventional manner by defining it as "the brilliance which you see the star", you as an observer always located at a constant distance of 32.57 years-light (10 parsecs). This last point is very important because if a star is far from us less than 32.57 years-light, its apparent magnitude will be greater than its absolute magnitude. And this is the case of our IUMMA whose distance from you is 144,371 years-light (apparent distance). Its apparent magnitude will be larger (around 12) to the absolute visual magnitude (around 14.3) as if she was seen at a greater distance (32.57 light years).

« Apparent visual magnitude: it will probably be reduced because of the interposition of a large cloud of cosmic dust which is at 3,682 parsecs, but it will be between 12 and 13 and therefore will not be visible for you but with photographic means.'

Our Oomomen visitors ascribe an absolute magnitude of 14.3 for the star IUMMA. Our Sun has an absolute magnitude of 4.8, which is a difference of 9.5. This means that our Sun is 7,900 times brighter than IUMMA. The apparent magnitude means "light really observed" and absolute magnitude means "light that would be observed if there were 10 parsecs (32.6 light years) between us and the observed object."

In the document D36, the IUMMA apparent visual magnitude is given between 12 and 13, which means a very low brightness. However the document D74, does not exclude the hypothesis that due to the dust cloud, the brightness could be extremely weakened: "But you would notice that the brilliant will be greatly attenuated because of the presence of a mass of cosmic dust which attenuates and reduced to an apparent magnitude of about 26."

Regarding the star IUMMA, if it had an absolute magnitude of 14.3 and an extremely low hypothetical apparent magnitude of 26, this would mean that the apparent distance would be about 71,390 light years. In this case the star IUMMA would be almost undetectable.

The Oomomen calculate the absolute distance from the Sun to IUMMA at 14.5 LY. They consult our various astronomical tables that mention that the distance from Wolf 424 to the Sun is 14.5 LY. So, they simply make the assumption that MAYBE IUMMA is Wolf 424, and it is nothing more than an assumption.

Even more to the point, the letter D21 tells us that « Unfortunately the errors made by you regarding the measurement of distances, are in many cases above 15% and in addition there are differences in the measures recorded in the various terrestrial stars catalogs. So it follows that it is impossible, even by a translation of axes performed with the utmost care, to identify any star codified by us with another cataloged by astronomers on Earth.'

This suggests to us that the distance criterion is a not reliable one for our research.

The upside of this is that the credibility-cautious approach of the Oomomen is justified. Indeed studies have shown that the characteristics of Wolf 424 are hardly compatible with the references concerning IUMMA. The Oomomen were not "wrong", they just made a conjecture. The documents are perfectly correct in this view. ULTIMATELY OUR VISITORS DO NOT TELL US WHAT THEIR EXACT LOCATION IS.

Like the Oomomen, we also conducted a first series of calculations and evaluations suggesting that "IUMMA" could be a star in our astronomical tables called "61 Cygni A" which is 11.43 LY away, while IUMMA is supposed to be 14.5 LY away. Then we received the letter NR21 giving us additional information. We made a second series of calculations. The distance criterion is not very reliable for our research. However the Oomomen initially thought Wolf 424 was IUMMA, so it seemed reasonable to think that Wolf 424 and IUMMA were still at distances that were not on opposite ends of the cosmos.

We also believed that the axis of Earth-IUMMA given in the documents was more reliable. We needed to be sure that variations of Ascension and Declination were also "reasonable". The letter D21 tells us:

RIGHT ASCENSION: 12 hours 31 minutes 14 seconds.

DECLINATION: 9 ° 18′ 7″ (area of the constellation Virgo)

The star XI Bootis B is located at the absolute distance of 21.85 LY, so it is within a "reasonable" range of Wolf 424. Moreover the axis of Earth—XI Bootis B is acceptable and consistent with the axis of Earth-IUMMA given initially. For XI Bootis B Ascension 14h 51mn 23s and Declination 19° 06′ 07″ are in an area similar to the constellation Virgo. The evaluation of the apparent distance depending on the brightness of XI Bootis B gives a distance of between 26 and 28 AL. Furthermore, as shown in the document D36, we have a apparent distance SUN-IUMMA between 27 and 29 AL, and the other another hand, an apparent distance SUN—XI Bootis B between 26 and 28 AL. These results are encouraging, so we continue. . .

The star XI Bootis B is an orange dwarf star, exactly as indicated in the letter D21 "our IUMMA is what you call a dwarf star."

The surface temperature of XI Bootis B is 4,600 K which is almost identical to 4580K of IUMMA!

Compared with the sun, the size of XI Bootis B is 71%, the brightness is 6%, and the mass is 74%, This translates to approximately 19,891 x 1030 kg x 0.74 = 1,471,934 x 1030 kg. The document NR21 gives us a value that is almost identical: "IOUMMA is a star of mass equal to 1.48 × 1030 kg".

In addition, radial variations in the orbit indicate the presence of one or more exoplanets. The critical data are given to us in the letter NR21: « Its light spectrum is redshifted compared to your Sun with photometric indices BV and UB equal to 1.15 depending on your reference system. »

However, stars with equal photometric indices UB and BV are very rare. Yet the values provided in the letter NR21 for IUMMA are precisely equal to the star XI Bootis B with photometric indices BV and UB equaling 1.15!

Star Name	Right Ascension	Declination (Virgo constellation)	U-V and B-V	Mass $(10^{30}$ kg)	Temperature	Spectral Type	Distance absolute LY
IUMMA	12:31 am' 14"	9 ° 18' 14"	U-B and B-V equal 1.15	1.48	4,580 K	K 3.5	14,421
Wolf 424	12:33 am' 22"	9 ° 01' 05" 9	U-B = 1.19 B-V = 1.84	0.151 and 0.245	2,000 /2,500	M 5.5	14.05/18
XI Bootis B	2:51 pm' 23"	19 ° 06' 07"	U-B and B-V equal 1.5	1,471,934	4,600 K	K 5V	21.85

Why did our friends think Wolf424 was IUMMA?

Are they so clueless about (terrestrial) astronomy that they mistook one star for another?

Or is there some astronomical criterion that led them astray?

Of course, if there's one thing our friends couldn't have been unaware of, it's that Wolf424 is a binary star, completely incompatible with a developed, habitable solar system. It has been well known to astronomers since 1941. Could they have confused a binary star with IUMMA, a single star?

No. So why this mistake?

One initial hypothesis is that IUMMA must be a binary star....

Xi Bootis is a binary star. If we look at the comparative table in Presence2, we see that all the fixed structural parameters of Xi Bootis B are the same as those of IUMMA, with very few discrepancies. However, its location and distance do not match.

We should also note that, subsequently, our friends never told us that it was the correct star. But they continued to provide clues to find it. As if they were testing our intellectual abilities…

ERRORS OR I.Q. TESTS?

When studying the Ummo documents some information or data seemed curious to us. However after careful study and verification calculations, we were always able to find a coherent explanation. Were we being subjected to tests of our I.Q.?

NR20 « We have deliberately omitted some information that you need to deduct yourself. »

FROM THE IQ TEST TO THE TRUE INVISIBLE COMPANION

The quote from Letter NR20 ("Know that we have sometimes intentionally slipped errors into our reports to assess your critical thinking skills…") serves here as a fundamental key to understanding. If we accept the idea that the Oummain network constantly gauges humanity's (and its scientists') ability to assimilate, verify, and question raw data, the apparent anomaly of the star Wolf 424 takes on an entirely different dimension. It is no longer a gross error in extraterrestrial astronomy, but a cognitive filter, a puzzle tailor-made to separate the naive reader from the rigorous researcher.

1. The Trap of Terrestrial Astronomical Tables

To understand the nature of this test, it is essential to consider the context of the first human interactions in the 1960s. At that time, terrestrial astronomy was rapidly evolving, but our parallax measurements (which allow us to estimate the distances of nearby stars) still had significant margins of error.

The Oummans know with absolute precision the position of their solar system relative to ours. They have estimated this three-dimensional geometric distance to be approximately 14.4 to 14.5 light-years. Knowing that a Norwegian ship transmitted a radio signal between February 5 and 7, 1934. This signal was then picked up on the planet Ummo "about 14 years later." Since radio waves (electromagnetic waves) travel at the speed of light, a travel time of about 14 years to cover a distance of about 14 light-years is perfectly logical and consistent.

With access to our scientific databases and star catalogs from that era, they searched for a star cataloged by humanity that corresponded to that distance along the appropriate line of sight. They then found Wolf 424.

Their wording in the famous D41 letter is thus marked by formidable semantic caution: they do not categorically assert that IUMMA is Wolf 424. They hypothesize that the star we call Wolf 424 likely corresponds to IUMMA, given human distance tables.

This is where the intellectual trap snaps shut. The Oummans were well aware of the binary and eruptive nature of Wolf 424. By designating this imperfect target, they issued us a challenge: "Will you stop at this inconsistency and reject our entire corpus, or will you use the other numerical data in our letters to correct the error yourselves and find our true star?"

2. *Cognitive Dissonance and Nominal Characteristics of IUMMA*

The contrast between Wolf 424 and the description of IUMMA in human documents is the primary tool of this discernment test.

Let us review the terrestrial observational facts regarding Wolf 424 (Gliese 473 AB):

- It is a binary system composed of two red dwarfs of spectral type M5.5Ve.

- Their respective masses are minuscule, estimated at around 0.14 solar masses each.

- These are so-called "eruptive" stars (UV Ceti-type variables), whose brightness can double or triple in a matter of minutes due to violent magnetic storms spewing deadly radiation.

- Their orbits are extremely eccentric and close together (with a period of about 16 years), precluding the existence of a stable planetary orbit within the habitable zone.

In light of this, let us outline the profile of the star IUMMA as formally detailed in the reports from the planet Ummo:

- Mass: Approximately 0.73 to 0.74 times the mass of our Sun.

- Surface temperature: Close to 4,500 Kelvin (making it an orange/yellow dwarf star, likely of spectral type K or late G).

- Stability: A main-sequence star, calm enough to allow the formation of a complex biosphere on a planet orbiting at a suitable distance.

- Photometry: Very specific color indices (U-B and B-V around 1.15), which is strictly incompatible with an eruptive M-type red dwarf.

The discrepancy is complete. From an astrophysical standpoint, it is impossible for an advanced civilization to mistake its own sun—an orange dwarf with 0.74 solar masses—for a chaotic pair of tiny red dwarfs. Therefore, if the nominal identification of Wolf 424 is a decoy (or a simple crude beacon), what reliable data allows us to solve the equation?

3. The Reference Axis as Absolute Truth

The resolution of the paradox lies in the method advocated by Denis Roger Denocla in *Présence 2*. While the star's name is a misleading hypothesis based on our own documentary shortcomings, the spatial axis provided by the Oummains is an absolute fact.

The letters provide extremely precise equatorial coordinates (Right Ascension 12h 31' 14", and a corrected Declination) that point to a specific region of the constellation Virgo. The correct scientific approach, the one required by the "IQ Test," consists of ignoring the semantic background noise ("Wolf 424") and aiming our instruments along this mathematical line of sight, at a theoretical distance of 14.4 light-years.

What the Oummains are telling us between the lines is: "Look in this precise direction. Do not trust the obvious stars twinkling in the foreground. Look for a 0.74 solar-mass, K-type star with a B-V index of 1.15."

This logic paves the way for the only physical hypothesis capable of reconciling the axis, the distance, and the characteristics: that of the Invisible Companion and the Absorbing Cloud.

4. The Occultation Model: The Cosmic Dust Cloud

How could a star with nearly 75% of the Sun's mass, located just 14.4 light-years away, escape detection by our optical telescopes? The answer is explicitly given in the Oummain corpus, though often buried beneath the geometric data: the existence of a mass of cosmic dust.

Interstellar space is not empty. It is traversed by clouds of cold gas and silicates (dark nebulae). The Oummains specify that a cluster of absorbing matter lies exactly in the line of sight between our solar system and IUMMA.

In terrestrial astrophysics, this phenomenon is well-documented under the name of interstellar extinction. Dust particles absorb and scatter visible light (particularly short wavelengths, such as blue and ultraviolet) . This phenomenon has two major consequences:

1.   Dimming: The star's apparent magnitude decreases drastically. A bright star located behind a dense cloud may appear to us as a very faint object, or even become completely invisible in the optical spectrum (visible light).

2.   Interstellar redshift: The star's light is shifted toward the red end of the spectrum.

Thus, the mystery is solved. The star IUMMA is indeed present, massive, and hot (4500 K). But the light it emits is absorbed by this curtain of dust. Its residual light, if it reaches us, is so faint that it has blended into the diffuse background or been overlooked by our historical records.

5. The Revelation of the Invisible Companion

Here we arrive at the brilliant synthesis of this cosmic investigation. Wolf 424 is not IUMMA. Xi Bootis, despite having fascinating structural characteristics and being very close to the target (as the comparative table shows), is neither on the correct spatial axis nor at the correct distance.

Geometric and astrophysical reality demands the presence of a third body: the true sun of Ummo.

This Invisible Companion does not necessarily orbit the two red dwarfs of Wolf 424. It is simply located in the same region of space, along our line of sight. The red dwarfs Wolf 424 A and B are most likely located in front of the dust cloud, or at its edge, which explains why they shine brightly enough to be detected by our observatories since 1941.

IUMMA, on the other hand, lies in the background, shrouded in its cocoon of light-absorbing dust. The Oummans used the visible beacon (Wolf 424) to draw our attention to this sector, knowing that it would take humanity several decades to develop the tools necessary to pierce the optical illusion.

6. Conclusion of the Analysis: A Challenge for the Future of Astronomy

The Invisible Companion hypothesis, coupled with the deliberate intelligence test, offers a revolutionary interpretation of the IUMMA paradox. It resolves the apparent contradictions and strengthens the internal consistency of the Oumma case. Far from being a gross error by incompetent observers, the identification of Wolf 424 is a psychological engineering maneuver.

This conclusion also formulates a testable scientific prediction. Since cosmic dust blocks visible light but allows far-infrared radiation to pass through (which corresponds to the heat emitted by the star or re-emitted by the dust), IUMMA should be detectable by ultra-sensitive thermal instruments.

The advent of next-generation space telescopes operating in the infrared spectrum (such as the James Webb Space Telescope) could very soon provide the definitive answer. By scrutinizing the exact axis at 12h 31' 14" Right Ascension, beyond the eruptive turmoil of the red dwarf stars in Wolf 424, our instruments may finally pierce the dark veil to discover the thermal signature of a K-type star with 0.74 solar masses.

The IQ test imposed by the Oummains would thus culminate in a human triumph: that of having managed to look beyond the obvious flaws in an erroneous astronomical table to deduce, through logic, mathematics, and astrophysics, the existence of an invisible celestial body that might be home to our elusive correspondents.

THE "INVISIBLE COMPANION" HYPOTHESIS: SOLVING THE MYSTERY OF IUMMA

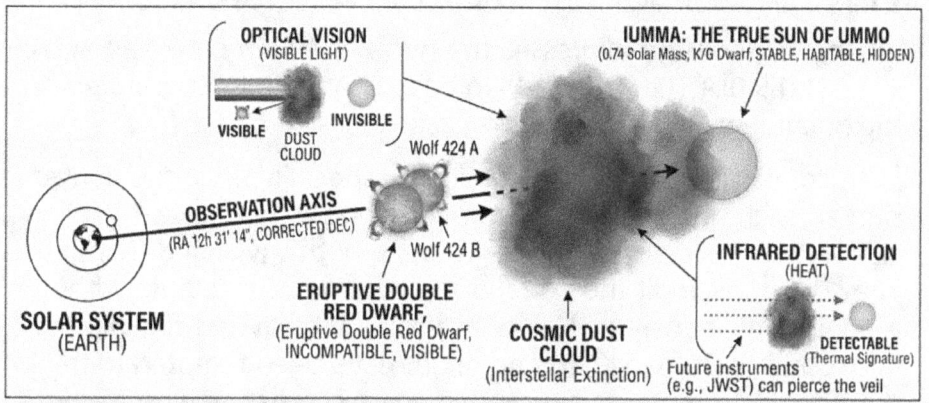

AN UMMAN I.Q. TEST: Use the precise coordinates (14.4 LY) and nominal characteristics (K-type) to look beyond the deceptive appearance of the two Wolf 424 dwarfs and find the invisible star hidden behind the dust, confirming its presence via infrared.

AN I.Q. TEST FOR COUNTING?

The first example I will give to illustrate my thinking, involves two errors that are almost impossible to make using a simple count table. However for a human, he would have to make the Ummo calculation in base 12 to find these two errors. Can you find these errors? (Cf. chapter Learn to count in Ummo language).

WUA EXPRESSIONS (MATHEMATICS) OF UMMO. We added the table of basic signs and as a result, we put other with the corresponding expression.

Like myself, you probably would not have identified the two errors at first glance, or even suspected any errors, if you had not made the effort to check the table yourself.

The terms express numbers on the left in our base 10 and to right the same number is spelled in base 12. There are 38 accurate numbers out of the 40. The 36th and 40th terms are wrong; they were noted with the terrestrial's numbers 99 and 120, which is incorrect.

12=	21=	29=	67=	91=
13=	22=	30=	68=	93=
14=	23=	31=	75=	96=
15=	24=	32=	77=	99=
16=	25=	33=	80=	100=
17=	26=	34=	82=	101=
18=	27=	35=	85=	105=
19=	28=	36=		120=
20=				

The 36th design expresses the number 82 in base 12, so (121 x 8) + (120 x 2) = 98 in base 10 and not 99 as stated in the document.

The 40th design expresses the number 100 in base 12 so (122 x 1) + (121 x 0) + (120 x 0) = 144 in base 10 and not 120 as stated on the document.

A conversion error seems unlikely, especially since the conversion algorithm is very easy to apply as long as one takes the exercise seriously. However, these errors are both trivial and harmless. They don't discredit the table, but they are undetectable without a minimum of analysis. However, if one actually did this analysis and found these two errors, he might have been tempted to think that the document was a forgery made by someone who was less competent than he looked. So that on closer observation, all of this looks very much like an educational assessment test to measure the reader's ability to manage the information.

AN I.Q. TEST IN BIOLOGY?

There is another example in the annex of the Alicia Araujo letter of an inversion that is not easily detectable in the first reading. In this document the Oomomen write:

(IXOUURAA)   Acido desoxirribonucleico (RNA)
(UCUORAA)   Acido ribonucleico (DNA)

This is an inversion between the English DNA and RNA acronyms. Again, this error is both trivial and inconsequential, but not readily detectable without a minimum of attention.

In addition still without questioning the principle of the text presented, we can see basic errors on the combination of DNA nucleotides in each drawing provided. These errors in the drawings are manifestly incompatible with the level and accuracy of the text itself. Why are these errors in the diagrams, but not in the main document? Are we again being tested by the Oomomen?

Estos componentes básicos están enlazados de la siguiente manera

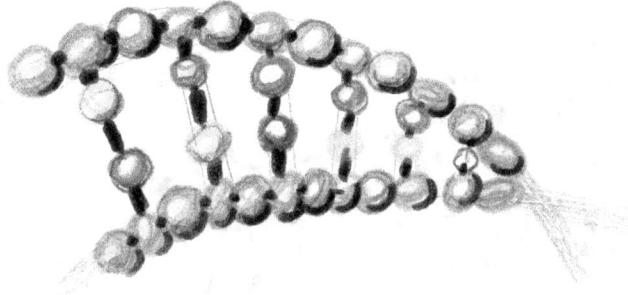

que forman una fracción elemental de la doble hélice citada.

Obsérvese que las situaciones relativas de los cuatro últimos componentes (Adenina) (Timina) (Citosina) y (Guanina) van alternándose de una manera aparentemente caprichosa, Pero es precisamente esa distribución de los cuatro IGOOMII (Factores simbólicos del CODIGO) la que permite portar el mensaje. Cada cuatro bandas o parejas ᗯᑎᗰ puede representar un símbolo IGOOA SK (GENE SK) programador de proteínas (este es uno de tantos símbolos genéticos).

AN I.Q. TEST IN LOGIC?

The same kind of intellectual test is found in the presentation of

a revolutionary engine running with a chemical reaction of Xenon

tetrafluoride which is quite unknown to us.

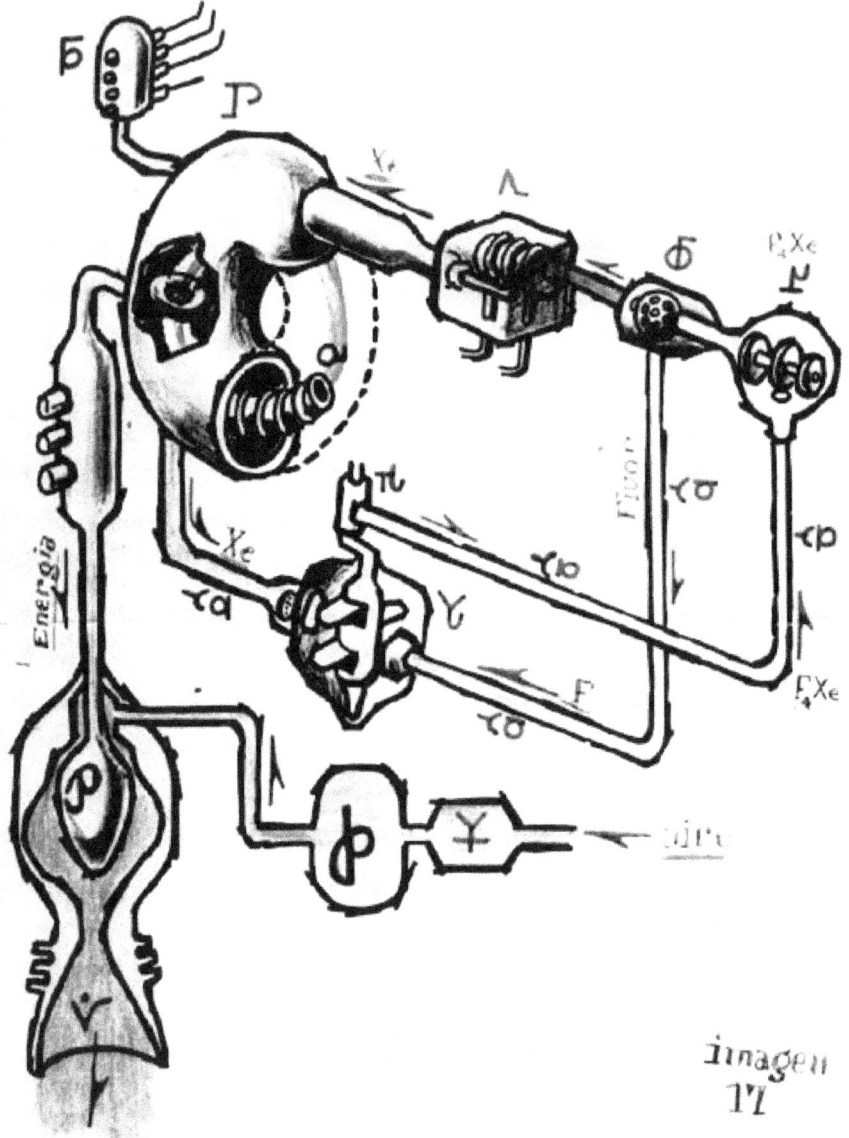

D41-6: A BUUTZ OF GOONNIAOADOO DESCRIPTION

The principle of this BUUTZ, however, is well-known on our planet. Xenon tetrafluoride is decomposed in the equipment ᚋ i.e.: the Xenon toroidal passes into the reactor already mentioned, while the Fluor is channeled into the regenerator ᚲᚢ by previously storing at high pressure in the tank ᚴ.

When the engine is stopped, the Xenon gas is recovered

through the line ᚱᚦ to be synthesized again in tetrafluoride ᚱᚦ.

The energy created by the plasma chamber $\mathcal{P}$ and this is where the air which was previously liquefied by the equipment $\mathcal{P}$ is channeled to the expansion chamber Υ and stored in the tank $\mathcal{O}$ expands violently and is projected downwards through the nozzle $\mathcal{V}$.

However when one observes the diagram, the circuit of fluorine return and xenon tetrafluoride in the picture is not consistent with the explanation. The circuits of fluorine and fluoride were reversed. Is this a small joke to see if you really followed the explanation?

AN I.Q. TEST IN ASTRONOMY

An Ummo document announced in 1979: "another planet located 7,898 kilometers beyond Pluto (average distance to your Sun)." [D116]

If you had no skill in astronomy, you would not have noticed that this is totally incoherent and stupid. 7,898 kilometers represent the approximate between Paris and New York.

No planet can be at this distance. Does this mean that this entire document is false? No. But if you have skills in astronomy you will pass the test successfully: the astronomical distances between two planets in our solar system are absolutely not measured in thousands kilometers, but in millions of kilometers. To be meaningful, the above quote must be read: "another planet located at 7,898 [millions] of kilometers beyond Pluto (average distance to your Sun)." [D116]

If you can brilliantly pass your test in astronomy, you will then have very interesting information to work with. (see Astronomical evidence)

CONCLUSION ON I.Q. TESTS

There are only a half-dozen errors identified on the 1400 pages of documents, which shows us that the Ummo documents have extremely reliable content.

The errors are very rare. They are only detectable if the reader has actually made the effort to VERIFY the document information. On analysis we find that these errors or omissions are easily rectifiable. Given their rarity, we cannot exclude the possibility that they are simply unwanted. However one might reasonably wonder whether the errors may have been deliberately inserted to test the reader and check his ability at analysis and critical thinking.

We can also wonder why, on the one hand, the Oomomen transmit information, yet, on the other hand, they ask us not to communicate with them? This information is for all Earthlings, so should it not be disseminated? Why should data of such great collective and social interest not be made known?

D1378: "We felt obliged to pay you back by giving you in return, information about our culture."

Example: " . . . steroid molecules to prevent rejection of heart transplants. . ."

NR20 17/01/04: "I advise you to keep this correspondence for yourself or to destroy it"

Could this not also be a major I.Q.. test? Are our visitors assessing our ability to take responsibility for disobeying and making social and collective issues our priority?

PREDICTIVE EVIDENCES

Certain information contained in the Ummo documents might have seemed completely false or imaginary in the 1960s. It was easy then to say that it was all a hoax.

But now the alleged false information is correct and can be verified 20, 30 or 40 years later. The so-called pranksters and a few "debunkers" have, in the end, been proven to have extraordinary scientific knowledge and predictive capabilities. . .

THE MYSTERY OF CROP CIRCLES

The Ummo documents refer to Crop Circles, these large and complex designs that are mysteriously made in the wheat fields since 1990 in U.K mainly. They talk in these terms:

"Crop circles that appear spontaneously in the middle of your fields come as a surprise to you. A great number of your brothers really believe that OEMMII (here terrestrial human) pranksters with mere wooden boards could have made them. When will such gullibility come to an end? Yes, these signs are drawn, in their great majority, by traveling OEMII, foreign to your planet. They are not a result of our actions, but we know the race of OEMMII who produces them. Their morality is high and we do not reprove of their actions. Their purpose is not merely to exercise a form of artistic expression, to the detriment of your crops, but to raise awareness progressively about the reality of extraterrestrial life, through a legitimate questioning about the origin of these signs. The deliberate disrepute, sponsored by State investigation services and relayed by information broadcasting organizations, will necessarily give way, beyond a certain threshold of credibility, which has turned out to be higher than mere logic would have led us to believe." NR17 (September 2003)

Based on these indications, i could show in 2005 (book published in 2007) that the Crop Circles were actually made by a single exocivilisation from a star with a complex planetary system located 27 LY from Earth. (See "PRESENCE UFOs, Crop Circles, and Exocivilizations.")

AN UNKNOWN COSMOLOGY

In 1966 the Oomomen's cosmological model described a universe composed of multiple pairs of cosmos/anti-cosmos having at least 10 dimensions. The first of our cosmological models that was really close to this concept appeared in 1998, 32 years later.

It is important to note that to date no cosmological model has developed concepts associated with angular dimensions like the Ummo model.

This universe model is somewhat similar to that described by physicists Igor Dmitrievich Novikov and Andrei Sakharov in 1970. It consists of multiple pairs of cosmos "leaves". But for Sakharov pairs of leaflets are successive in time, while for Oomomen the "leaves" are "simultaneous". The first models which were truly close to the Ummo cosmological model were called "brane cosmos" and traced to the work of Lisa Randall and Raman Sundrum in 1999.

They in turn were inspired by the work of Arkhani-Hamed, Dimopoulos, and Dvali in 1998. Again we must note that to date absolutely no current cosmological model develops concepts asso-ciated with angular dimensions, such as those outlined in the Ummo cosmological model. To somewhat understand this model we must imagine that machines can move from one end of our cosmos to another by taking a shortcut which uses another 3D frame in our cos-mos. I explain this in detail in the book "PRESENCE UFOs, Crop Circles, and Exocivilizations."

A CHEMICAL PRODUCT PIONEERING

Another argument for authenticity is that all of the topics cove-red in the UMMO documents are more advanced than their time. For example, in 1966 the description of an environmentally friendly engine which operates by using a chemical reaction of xenon tetra-fluoride (see reference D41-6) would have required that its writer be part of a small crew of research chemists who had followed Neil Bartlett's 1963 discovery of this chemical reaction at the University of British Columbia. If the letter describing such an engine in the year 1966 came from Planet Earth, it could only have been written by a select person from a very elite group. If that is not enough, to be consistent with the rest of the text, the writer would also have had to have huge skills in many other areas. . .

Xenon tetrafluoride

LIFE ON MARS PREDICTED...

Ever since the XVIIth century it has been politically correct to think that Mars is inhabited by some form of life. At the end of the XIVth century Giovanni Schiaparelli observed something he called "channels" on the planet Mars. Research on the Martian channels was made popular in the XXth century by Percival Lowell and Earl C. Slipher, who took numerous photographs of these "channels" from that date and up until the 1960s, looking for life. In 1965 detailed snapshots from the vehicle Mariner 4 showed that these "channels" were actually great canyons caused by erosion.

However, we now know for sure that there are unicellular organisms on Mars, because they have produced organic gases such as methane, which attests to the fact that there was past bio-organic activity.

But in the 1960s nobody knew what would be found on Mars. Yet the Ummo document announced the presence of these same unicellular organisms on Mars in the year 1967:

D57-2 01/30/67: " . . . not only forms protein and amino acids, but also of unicellular and multicellular plants simple."

A prediction comes true after 30 years...Another prediction in a document referenced D46 (from 1966) gives startling news:

«... We caution you that the whole series of antibiotics that you have made are now helping to create in the future of new strains of viruses and pathogens much more resistant and immunized against your pharmacological, and therefore, in a future of 180 years + or - 10%, 72% of species you cataloged will be as virulent as before the onset of these drugs."

STEM CELLS WAY BEFORE THEIR TIME...

There were also other newsworthy subjects which were only being explored by specialists in hematology and genetic therapy in the mid 1960s. However, even though hematopoietic stem cells were identified in 1932 by Paul Murray, they had never been the subject of genetic manipulation.

The first clinical trials of gene therapy (to replace the simple marrow transplants in cases of genetic blood diseases) were conducted between March 1999 and May 2002. Yet in the year 1966, the document D47-3 clearly suggests that we use genetic manipulation on cells:

«... We can transform the nucleus cell of any tissue in various ways. That is to say that we can generate real ARTIFICIAL ATYPICAL CELL [. . .]. What is actually achieved in practice is to change the nature of a cell. Imagine for example an area overrun by FIBROIDS, connective tissue cells. Well, acting on the chromosomes of the nucleus it is possible to convert one or thousands of such cells into cells such as nerve cells, that is to say NEURONS whose structure is completely different."

.

THE IMPOSSIBLE ANALYSIS

The following letter D108 was received on 21/04/1973, and it presents a biological experiment:

"A few months ago (the project had been launched in an old cottage near Marseilles, France) my brothers had developed a research program related to an unknown viral entity on the planet Earth but familiar (although limited) of UMMO brothers biologists. The structure of these viral specimens has a certain similarity with chain corresponding to the circular DNA virus known on EARTH like 'rat poliomavirus' due to its spatial arrangement but this entity was provided with a capsule much more complex."

This is all very interesting because according virology experts the virus SV40 has been widely used since the 1970s. However the "rat polyomavirus" is a pathogen that has hardly been studied to date, and was not studied at all in 1973. . .

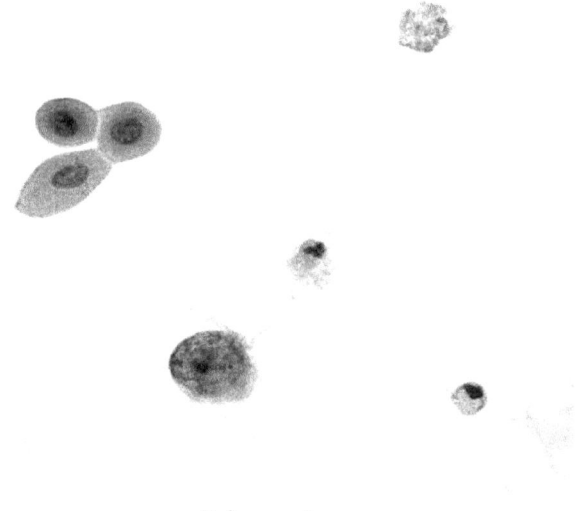

Polyomavirus

In addition, in order to analyze the DNA of the hull of "rat polyo-mavirus" in 1973 you would need to be in one of the very few labo-ratories that was capable at that time of performing long, delicate, and expensive operations.

So to compare the DNA of a virus with the one of "rat polyomavi-rus" in the year 1973 on the planet Earth was quite simply impossible.

"But we also find that this protection [immune-protection] ceased shortly before the histologic tissue degeneration, by the animal's death or by necrosis of the cell area. Under these condi-tions the cells cease to be immune-competent."

The above-quoted phenomenon is currently incomprehensible to any virologist on Earth.

A STRANGE KNOWLEDGE OF GERMANIUM

The D41-3 in 1966 letter indicates that Oomomen use compounds of, which have physicochemical properties of light transmittance on UMMO. This is to say pretty much the same as on Earth from red to purple.

"It is 196 UIW all the XAABI is silent. That night, the panels sound insulation built with a compound of germanium totally transparent and that we use roofs were closed. Not a single window communi-cates with the outside. Only a trained ear can hear the whistle that produces very low air properly assayed for temperature, humidity, ionization and ozonation, air is expelled through the nozzles of each IAXAABI (home)."

On Earth, this knowledge of the physico-chemical properties of transparency to visible light of compounds based on Germanium is known by a few experts whose communications are very limited and confidential. Ge oxide (GeO2) will be used later for military equipment infrared detection for wide-angle lenses or optical microscopy, but also as an element of optical fibers.

"The crucial attenuation limit of 20 dB/km was first achieved in 1970, by researchers Robert D. Maurer, Donald Keck, Peter C. Schultz, and Frank Zimar working for American glass maker Corning Glass Works, now Corning Incorporated. They demonstrated a fiber with

17 dB/km attenuation by doping silica glass with titanium. A few years later they produced a fiber with only 4 dB/km attenuation using germanium dioxide as the core dopant. Such low attenuation ushered in optical fiber telecommunication. In 1981, General Electric produced fused quartz ingots that could be drawn into fiber optic strands 25 miles (40 km) long."

We also note that Germanium is a rare and expensive element, its concentration in the earth's crust is very low. While on UMMO it is an industrial product:

D1378 " . . . the transmutation of chemical elements is not a problem. We can produce with great economy of resources, not only protein but any molecular weight (iron, titanium, cesium, Pentane, Haemoglobin, complex derivatives of silicon and germanium. . .) from raw materials as simple as water, oxygen, or the sand."

D 41-4 ". . . Similarly, other non-food consumption but topical in the homes of our planet such as acids, liquid propane, silicone polymers, zirconium oxide, germanium compounds are compressed into a cylinder (if they are powdery or solid) or poured into cylindrical containers if they are liquid. Distributing centers exists on UMMO linked by an extensive network of underground pipes NUUDAIAA or conduction at all XAABII (HOUSES) in the region. . ."

In 1966 the Oomomen know in detail and perfection all the physico-chemical properties of Germanium:

D58-3 "Even if the number of life forms consistent is very high, we warn you that we have checked that some life forms ARE ABSOLUTELY NOT POSSIBLE. For example suggests that in the Stars inhabited would have developed living beings based on different biochemical basis has no foundation. Eg based on the chemistry of silicon or germanium Chemistry."

Meanwhile in 1966, on Earth only a handful of experts just beginning to discover these physicochemical properties . . . and many others are still to discover...Anti-graft rejection molecules unknown

In the early XXIst century, organ transplants still face the problem of transplant rejections. However a Ummo document from the 1970s explains that the use of steroid molecules would have the transplanted avoid these graft rejections.

These steroid molecules are products largely unknown to terrestrial biologists. Although they are referenced in the databases of pharmacological products, no study has ever been performed on them. The effects and properties of steroid molecules are unknown. However, I asked a biochemist to enumerate the biochemical processes that lead to the production of these molecules. His words are paraphrased below...The steroid compound known by the Oomomen as AUMAOYEE results from the conversion of pregnenolone to progesterone in two steps

1) a redox 3 beta 2) followed by an isomerase which moves the double bond from one kernel onto another.

The AUMAOYEE molecule is considered to be an intermediate. The step that leads to the progesterone is the oxidation of AUMOYEE. The other steroid compound called IMMAOYEE may result from hydroxylation of the compound AUMAOYEE. Then IMMAOYEE can turn either into cortisol or hydrocortisone.

Will there ever be a way that we on Earth can use these steroid molecules to prevent graft rejection?

Preliminary study redacted by DENOCLA

THE NAWEE EVIDENCE

The Spanish letter D45 of 1966, states:

"GOSEEE: cosmic unit length used on UMMO, it is equivalent to the distance of IUMMA to NAUEE at its peak 76 × 126 ENMOO)"

So 76 × 126 ENMOO = 424,328 km. If such a planet were to move some 425,000 km from its star, it would be immediately charred to ashes, and become part of the star. This would happen whether it was 425,000 kilometers from the center, or from the periphery of the star, because with the solar wind, the orbit would decay and the planet would fall into the sun in less than a million years. So it seems that the indications in the letter D45 are wrong. So should we then conclude that the entire Ummo file is a fake? The characteristics and data of the (NAUEE) NAWEE planet were given again 43 years after the letter NR21 from 13/03/2009:

Equatorial radius	$R = 6.22 \times 10^7$ m
Weight	$m = 9.91 \times 10^{25}$ kg
Orbital eccentricity	0.026
Average distance to IOUMMA	5.97×10^{10} m
	(59.7 million km)

The value given in the NR21 for aphelion (apogee) of NAWEE is 59.7 × 1,026 = 61,250,000 million km.

This fits perfectly with our mass data and it is otherwise very similar to our astronomical calculations. Thus 43 years later the letter NR21 has highlighted the fact that the letter D45 contains false data. Should we then conclude that the document D45 is a fake and that the document NR21 is the only authentic document?

It seems more logical to conclude that by the year 2009 the writer of the NR21 letter had become aware of the distance error in the D45 document referring to NAWEE. One must wonder why he was interested enough to give us the corrected and accurate information 43 years later.

If we explore more closely, the letter D45 was written in 1966, at a time when the Oomomen used Earthlings typists. If we substitute the exponent 6 for the exponent 8, we can now see that the calculations are perfectly accurate. The mistake came from the 1966 typist, who typed a 6 instead of an 8:

. 76 x 126 ENMOO = 424 328 km.

. 76 x 128 ENMOO = 32.68 millions d'ENMOO
= 61,23 millions de km.

The two numbers of the D45 and NR21 match at 0.00004, so the two documents are completely accurate considering the 43-year intervals. Moreover, for complex astronomical reasons, the NAWEE distance is exactly the value which aligns with the concept of XEE (Ummo year) and is consequentially the interval between 2 triple alignments UMMO-IUMMA-NAWEE. Do you believe that a prankster would perform such a prank?

ASTRONOMICAL EVIDENCE

To complete our evidence, the "cherry on top" was the 1979 prediction of the existence of ERIS (2003 UB313 Xena), which came nearly 25 years before the planet would be discovered in 2003.

[D116] May 23, 1979: " ... another planet beyond Pluto. .."

The mere indication of the presence of an unobserved planet after centuries of astronomical observations, and 25 years before the official discovery of that planet is similar, for its own time, to the Newtonian discovery of the location of Neptune. This information alone is sufficient to demonstrate the authenticity of the Ummo documents.

THE IMPOSSIBLE PREDICTION IS REAL!

The truth is that despite decades of highly instrumental astronomy and telescopes capable of diving into the heart of the cosmos, we were not able to detect the planet ERIS (2003 UB313 Xena) until the year 2003. Could one possibly imagine that the authors of the UMMO documents could have guessed at the existence of this planet? If so, what would be the probability that such a prediction would come true?

So the "cream on the cherry", one might say, is that the D116 document dated Mat 23, 1979 predicts not only the planet Eris, but also its mean orbit.

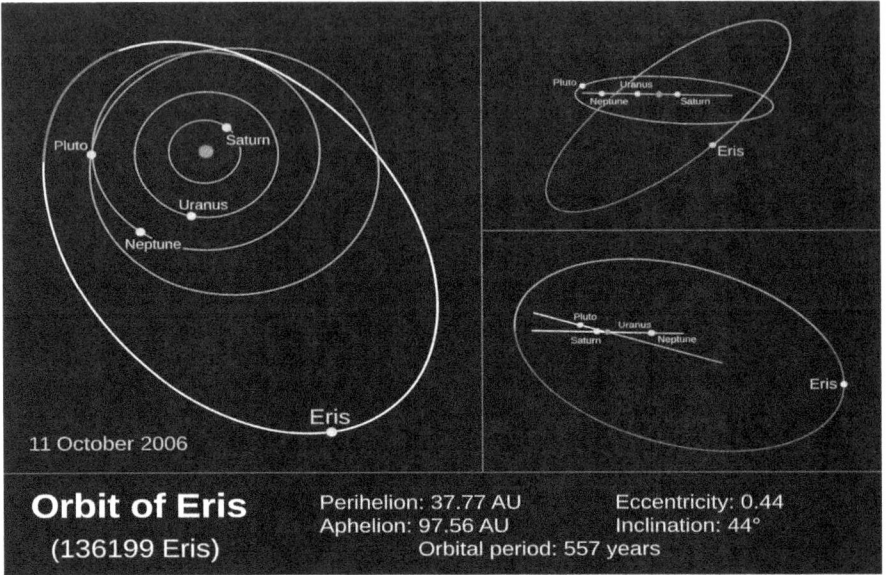

Orbit of Eris

(136199 Eris)

Perihelion: 37.77 AU
Aphelion: 97.56 AU
Orbital period: 557 years

Eccentricity: 0.44
Inclination: 44°

11 October 2006

The average planetary orbit is on an atypical ecliptic of about 14,518 Mkm. The D116 document indicates this very precisely:

[D116] May 23, 1979: "another planet located at 7,898 [millions] of kilometers beyond Pluto (average distance to your Sun)."

The average distance between Pluto and the Sun is 5,906 Mkm. 7,898 Mkm + 5,906 Mkm = 13,804 Mkm; the margin of error between the two orbits is thereby placed at 5%, which is ridiculously low.

What great anonymous genius could have predicted these results with such precision? This result is not understandable to the ordinary statistician or astronomer—only the brilliantly crazy ones.

ONE OF THE GREATEST REVOLUTION FOR TERRESTRIAL HUMANITY

To determine the existence and the orbit of ERIS (2003 UB313 Xena) 25 years ahead of its discovery is ABSOLUTELY IMPOSSIBLE. This is one of the many proofs of the authenticity of the file UMMO.

This is also the evidence of the presence of exocivilizations on Earth whose purpose is to raise our level of knowledge, and to develop our awareness and responsibility for our role in the cosmos. Each person on the planet Earth must now begin to take an active part in the construction of a new terrestrial humanity; humanity which bases itself in universal cosmic ethics of all humans being responsible and caring on earth and also in this cosmos.

This is the biggest revolution for terrestrial humanity.

The extraordinary details of the UMMO file are verified 20, 30, and 40 years later and show that the individuals who wrote the letters had extraordinary scientific knowledge and predictive capabilities which would be impossible to achieve if they did not already have the information themselves at the time of the writing.

In the next section we will see that the deciphering of the Ummo language allows us to make even more discoveries, just as stunning and surprising...

*

Exploring the UMMO Language

The 1400 pages of known Ummo documents contain unknown "words" which have been resistant to analysis and many decoding attempts for half a century.

—You also mentioned the possibility that the "words" of the Ummo language that are contained in the documents are the key to the authentication of the documents. Could you explain that?

The first point is that the Ummo "words" can only be written in phonetics because their original scripts as they are presented in the documents are based on ideograms of conventional correspondence, that are probably conceptual. Without any clues as to the correspondence between an ideogram and its conventional value, it would be impossible to do the decoding. Because of these clues, phonetic writing of an Ummo "word" is possible. Because of the strong equivalence between Spanish spelling and phonics, Spain was the one chosen to initiate the experience of written communication with Earthlings. So those were the terms of the earth experiment.

The second point is the authentication of the documents themselves. They are dispersed to multiple recipients. The deciphering of the Ummo "words" requires the gathering of many letters together. In order to "break" the code we had to have a collaborative action of collecting the documents. Ultimately, when we had gathered enough material we could analyze and authenticate the documents, and therefore identify the discreet presence of Oomomen on Earth. So it seems that the documents were voluntarily distributed along with a "key". And it was up to Earthlings to find that key.

Idéogram	Phonetic	Definition	Idéogram	Phonetic	Definition
ᚻ	ENMOO	Ummo unit of length	⊕	WAALI	Statistical diameter of our galaxy
ᚾ	UIW	Unit of time on UMMO	ᙁ	?	Speed of light
gŧλ	INOWI	Fruit	⌐∞	GOSEEE	Cosmic unit of length
Ϥ	BIEYAEYEODOO	Chains of polypeptids	ᚢ	XANMOUULAYA	Central computer
Ƴ	ESEE OA	Conscience	Ⴔ	XAN ELOOWA	Control screen

— You worked for several years on these unknown "words" in these documents. Does this work bring additional guidance?

The 250 known documents number over 1,400 pages and make use of tetravalent logic. Tetravalent logic is applied consistently to all of the topics covered in the documents. It is in the letters of logic and mathematics, and also in the impressive corpus of cosmology. It is in presentations of new physical paradigms, and even in the codified language contained in the documents.

— What is so special about this tetravalent logic?

There are two particular points. The first is that on our planet that type of logic was common in ancient times in Europe before Aristotle in IVth century BC. It was also prevalent in Asia, and particularly in the Buddhist culture, where it was known as the tetra-lemne.

The second specific point about tetravalent logic is the fact that modern mathematics and all of our technologies have been developed solely on the basis of Aristotelian logic. That is to say that our logic stems from bivalent binary logic which is based on True and the False, and which follows the principle of the excluded. Since those ancient times there has been no development of tetra-valence in mathematical classic form until 2002–2003, when it was used by Alban Nanty and Norman Molhant. So all of these developments are new and very recent.

—So the topics covered in the documents, the concepts which are developed and this tetravalent logic, are uncommon and perfectly consistent. Are there other elements of coherence?

Yes, we can also note the numbering system used by the Oomomen. This numbering system has many features that are in common with the terrestrial numbering system, while also being totally specific and unique. The symbolic representation of these numbers is associated with a structured building having numbers from right to left, similar to our current numbers of Indian origin. The great similarity of these two numbering systems is both astonishing and yet also almost inevitable due to the fact that the means of developing complex systems of operations eventually leaves few opportunities for the composition of the basis structure of the numbering systems. Apart from the symbols of numbers 0, 1, 2, 3, 4, 8 and 12 that can be found in the very different cultures of the ancestral terrestrials, the other figures are simply graphics that have not yet been released. The numbering system used by the Oomomen has an amazing historical and anthropological coherence and an extraordinary semantic consistency with the conceptual graphics: the graphics of the figures correspond to the meaning of the concepts embedded in their word.

These amazing consistency on all these scopes are sufficient to betray an intelligence that goes far beyond the mere achievement of an anecdotal numbering system.

—For you, this amazing consistency of all these subjects shows signs of a culturally homogeneous group. Of precisely what nature is the language that is used?

These people have indeed used a very singular language. The corpus contains about 7,500 occurrences of "words" claimed to be from the extraterrestrial civilization from UMMO. The decoding efforts of various universities Spanish linguists and the many studies that followed were fruitless. In 2002 Jean Pollion did a line of research which he called "Ummo: real aliens". Then between 2003 and 2005 I myself succeeded in deciphering the language that is presented in this book.

While testing my predecessor's assertion, I was led to analyze everything in detail. For example, the concept of "atomicity" that he had detected was actually much more complex than he realized. Embedded in the "Words" was an overlapping hierarchical structure which was based on concepts that were strictly phonetic. This discovery enabled me to finally determine an effective reading methodology for these enigmatic "words".

It turned out that the limitation of our understanding was due to the fact that a second language level was superimposed on the first, and it is then complete the meaning more precisely. We do not have enough material right now to decode the second language level, but we do know that it will finalize the meaning of "words" of the first level. Ever since 2003, I have been writing a complete study in my "Research notebook", which has been published for free on my Web site. Also with a small team of volunteers, I have posted a database of "words" with their in-depth semantic analyses.

My work shows that all of documents belonging to the corpus for the past 40 years have a structure to the words in the language that is intrinsically identical.

— So you found the intrinsic structure of the Ummo "words". Could you describe it simply?

From the most recent documents to those dated forty years earlier, all are composed of a sequence of primary conceptual phonemes. Also, they are nested from major to minor, like Russian dolls. But some "insightful" minds would probably think this again was a KGB trick. Because the coding of the "words" is unique to the Earth as follows:

. phonetic (with Latin letters)
. conceptual
. functional
. by hierarchical nesting of primary concepts

— Could it be said that the Ummo language is something like the Chinese language?

That is a good question. And in 2000 Godelieve Van Overmeire also looked to find the similarities. He found that there was a vague similarity between Chinese and the Ummo language, but it was very vague because the only common point was the use of concepts to form the words. Further analysis by the sinologist Johannes Gehrs—in an article dated December, 2001 known as "Inforespace

No. 103" and entitled "Is the Language of Ummites Chinese? "— has clearly shown that the Ummo "words" were not taken from the Chinese language. So the answer is categorically no. I also confirmed this fact by showing that this language is based on phonetic concepts which are hierarchically nested from primary major to minor. This is absolutely nothing like the Chinese language, and to my knowledge there is no Earth language like it.

— So these "words" have a unique intrinsic structure that is consistent. Is there also a consistency between the content of the documents?

Yes, what I found to be amazing while reading these thousands of pages is that everything seems extremely coherent. And this consistency is not superficial. The ontological position of the primary concepts "E" and "O" clarify precisely the notions of "concept" and "entity". Two concepts necessary and consistent with the cosmological concepts, on the one hand, and with tetravalent logic, on the other hand. I explain these points in detail in the chapter entitled "Ontological positioning of the primary concepts".

— What would be the point of making up such a language?

Imagine a group of people using tetravalent logic for forty years and inventing from scratch a language with such a unique structure that it took the entire 40 years to be deciphered. Imagine too their handiwork being fully consistent with tetravalent logic and with a cosmological model which explains interstellar travel. Imagine too that the entire corpus is fully consistent with itself. And all this would need to have been done without anyone being able to say who did it or why!

Of course you must also have noticed the amount of work behind the implementation of such a corpus. Its magnitude is truly extraordinary and daunting but it lacks any identified goal.

This book demonstrates that the authors of these documents were not terrestrial researchers, but rather they were the aliens themselves. The documents allow us to enrich our knowledge and give us a way to achieve a certain level of understanding of the conceptual and functional language of these beings. In truth, it is quite difficult to understand the precise meaning of each conceptual and functional "word" in the documents because our minds are accustomed to our languages, where words mainly refer to objects, or concepts of objects. Their "words" are phonetic and they are compounded with primary concepts; the meaning of any given "word" comes from the interweaving of these successive primary concepts and from following a functional logic. Coherence of ideas, logic, and words emerge strikingly from the corpus, once we have understood the rules governing the reading of these words. This is one of the main keys to this issue. And this advance in the field of language contributes to the understanding of a paradigm of the universe which is radically new to this third millennium.

— Do you think that these documents can be utilized for scientific purposes?

Yes and no. I say no because even though the concepts included are exceptional, nothing is said in the documents about how to implement them, or about the practical means and immediate applications of the concepts. And I say yes because a significant number of the ideas presented in the 1960s documents have actually come to fruition. For instance, the CD-ROM burner, which is similar to an idea presented in the D66 document dated 1967, has now become a part of our everyday lives...

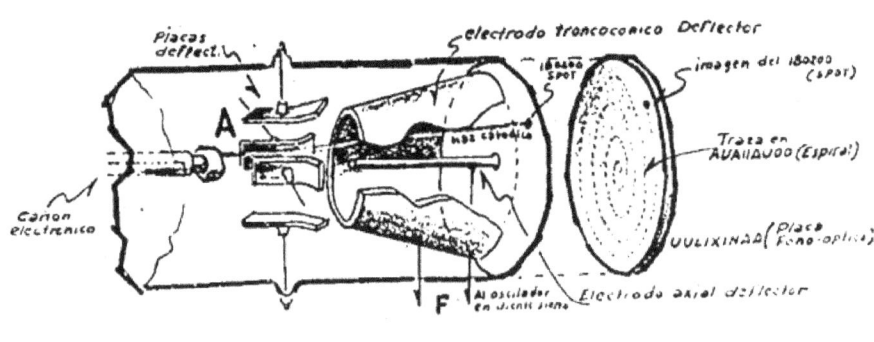

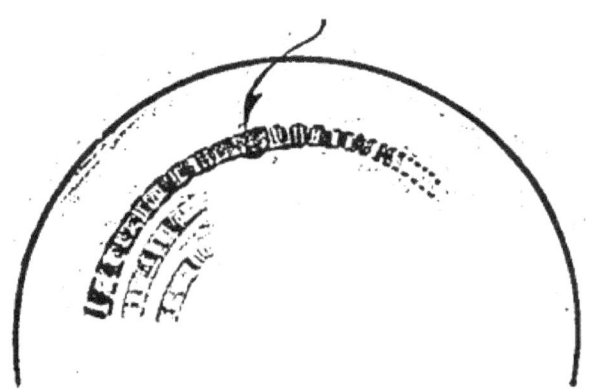

A kind of a CD-ROM burner disclosed in 1967—ref D66

— Do you think that UFO phenomena, and the Ummo file in particular, presents an underlying challenge to global politics?

Yes, the time has come for all people of the Earth to act on this new information.

STATE OF THE ART

About forty years after the first transmission/reception of the Ummo texts to Spain, this book explains in detail the logic of the Ummo "words". This explanation is based on the results of the semantic analysis of thousands of word occurrences that were analyzed by cross-comparisons, and it represents approximately 90% of all of the known texts and words to date.

The semantic analysis aims to clarify the meaning of words in order to improve our understanding of these detailed documents. The scope of these explanations concerns the first level of the language, while also acknowledging that there are two other levels. We will not attempt to explain the two other levels here, because they are currently beyond our reach. This first-level language "DU-OI-OIYOO (can be translated by Linking Language) uses [...] groups of related or connected phonemes that represent concepts, values and concrete objects and even ordered complex ideas. (D 77: OUR LANGUAGE AND LOGIC»

Since its first introduction in the year 1966, the Ummo language has been a mystery. The initial examinations of the language were totally unsuccessful. Then Ignacio Darnaude carried out the important work of compiling the documents, which would pave the way for other studies to follow. The first glimmer of understanding would not occur until the year 2002. To give a little background, in 1879 mathematics researcher Gottlob Frege penned something called "A Conceptual Writing". His work was followed by the similar work of mathematician and Nobel Prize Winner for Literature, Bertrand Russell. Mr. Russell was associated with Alfred North Whitehead who continued his work in 1913 in a piece called "Principia Mathematica". All of these people tried to show that logic creates mathematics. Then in 1918 in what he called "Philosophy of Logical Atomism" Bertrand Russell further developed his approach by conducting "deep studies on semantics (D87)". As indicated by Michel Seymour, Russell tried to show that analyzed expression can be decomposed into simpler expressions belonging to an extra-linguistic reality which goes directly into our intellectual or sense-based intuition, said another way the words could have a function as well as a concept. Although none of these works details the notion of primary concepts, they nevertheless contained the first clues to the basis of the logic behind a language that is conceptual as well as functional. This work was later continued by Ludwig Wittgenstein up until the year 1953.

Bertrand Russell, philosopher and logician

In 2002, following the Oomomen's indications, Jean Pollion identified that Bertrand Russell's concepts and approach showed that there was an underlying atomicity inherent in the language, meaning that the words had levels of indivisible units. The language of the first level type was what I would call a "Russellian" language which means that the words are conceptual representations of compounds and basic functions (cf. the semantic analysis of DU-OI-OIYOO). This corresponded well to the fact that the Oomomen indicated "groups of related or connected phonemes (esp: Voces) values and concrete objects and even ordered complex ideas".

In 2003, even though Jean Pollion thought that each spelling should be associated with a different word meaning, I was able to show that the principle of atomicity also applied to the primary concepts that were only phonetic. This atomicity had the effect of creating ambiguous results that were unusable.

However in 2004, I showed that the primary phonetic concepts had a nested hierarchical structure in which was embedded with the key for deciphering the "words".

Finally in 2005, I completed the determination of the phonetic values of the primary concepts and ultimately achieved the deciphering of the Ummo "words".

THE INITIAL RESEARCH OBJECTIVES

I was led to this research by "chance and necessity." But the practical result is that it is now easy for me to conclude that Man is not alone in the solidarity of the vast collective Cosmos.

My research on language began with the idea to reverse or consolidate various assumptions (which I will present the next book 'PRESENCE 3'.) I thought that if the words were composed of primary concepts, we must by necessity analyze the words to find the details that were not explained in the text.

To do this, it was necessary to decode words in a methodical manner that was also clear and reliable. Hence during the summer of 2003 I developed the first decoding method based on words reading with the use of the relation "get" between the letters of the words. Suddenly the deciphering "worked better". I then found Jean Pollion's first and most obvious error in the primary definition of the concept "I". By the end of 2003 I formally concluded, as many others had also sensed that the words were phonetic. I realized then that the work of my predecessor could not provide answers to questions that arose. (Such is the life of a researcher.)

So I worked on a new method from scratch, which stemmed from the gross principle of "atomic concepts", and was based this time on the phonetic approach. Next, after analyzing the words database, I was led to reconsider another five primary concepts. Finally, with this new version of the generalized semantic analysis another step was taken and the foundation was now established. Although some words are "extremely synthetic, [. . .], which make very complicate decoding them. . .", my initial research objectives were achieved. It was possible to read censored "words", to get a clearer understanding of certain scientific concepts, and to see the match between the "words" and what was described in the texts. And in some cases, the deciphering allowed me to discover things heretofore unknown . . . (cf. How the exocivilization's words analyze allows revolutionary discoveries)

FROM THE UMMO TEXTS

D21: "LANGUAGE AND FEATURES

The words noted in this document are approximate graphic expressions of their real voice."

D 77: "OUR LANGUAGE AND LOGIC

Our first objective of our thought was to draw up basic dialectical logic that was independent of the language. This was of vital importance when you consider the fact that our form of expression is bisynchrovalente and codification of verbal thought in two expression modes that can be phonetically simultaneous (one by a linguoguttural mechanism similar to the languages of Earth, and the other with a code that involves sequential patterns of phonemes)"

"The first language, DU-OI-OIYOO, can be translated by Linking Language and uses ideograms in their graphic expression and phonemes groups related or connected representing concepts, values and concrete objects and even complex ideas ordered. This is a vehicle used to converse with routine (home language, technical, macrosocial popularized). '.

NOTE 4: "... That kind of language not uses "words": the proposals encode components of the sentence by agglutinating (subject, verb and predicate as you would say) in the form of coded proposal."

"Thus the proposition "this greenish planet seems to float in space", would be expressed in our topical language (DU-OI-OIYOO) as follows: AYIIO NOOXOEOOYAA DOEE USGIGIIAM"

Ref. document 104-1: "Those quoted paragraphs are a literal decoding, taken as closely as possible the original report. The accuracy of the language version you are familiar with, gets along with the grammatical and semantic additions that make it intelligible, as our texts are highly synthetic without morphology spelling that is familiar to you, which makes them very complicated decoding without prior addition of verb forms, adjective, etc. "

D 541:" Understanding our language is difficult for you because normally we overlay within the same phonemes set, two simultaneous streams of speech. The modulation of these sounds and the repetition of phonemes are not redundant but it is distinct from the ideas expressed by the pronunciation of words and their order. "

D 41-1: "We will try by all means to write the acoustic picture of our expressions in Spanish spelling, although in most cases our phonemes can be made by several graphic expressions."

Automatic translation

The spelling and vocabulary in the Ummo documents have a much higher quality than the average level of any human text, either in French or Spanish. This implies that its writers have a deep knowledge of different languages.

In 1968 the United States Military created the first automatic translator to translate Russian into English. It was only used in a restricted context, so it is surprising that even two years before that our outer space friends seem to use automatic translation for French and Spanish translations, and also for some 10 other languages listed in the document. Moreover, the automatic translation of any of the Ummo document has fewer errors than any other text picked at random.

So try it yourself. Use any automatic translator to translate something and compare the results to any Ummo text.

The physiological context

My belief is that the closeness of the timbre of the Spanish language with the Ummo language, as well as the fact that the Spanish spelling is almost phonetic, led the Ummo to use Spanish for the first written communication experiment.

D108: "Three of our brothers among the displaced of UMMO on Earth at that time had specialized during their stay in France to the study of comparative phonetics Latin root languages and could speak with an aptitude as acceptable for languages such Romanian, Spanish, Catalan, Sephardic and Italian.

Unfortunately one of them had (as is quite normal among OEMII of UMMO) atrophied vocal cords which were an additional risk to overcome although our sound amplification devices make it unrecognizable as a default by a weak human observer."

Initially, this communication was done through various Spanish typists. The Oomomen had therefore implemented the first level of their language as words written phonetically. To write these words, they respected the structural principle of their functional language by associating each phoneme of their words to the nearest Spanish phoneme. So they created especially for us a kind of "meta-language" in Latin letters, which respected the original structure of their language.

If we follow the information we have, we see that the Oomomen's phonation is different from ours because of its comparatively small amplitude of bandwidth. The Oomomen compensate for this issue by having more sensitive hearing. They do not use genetic modification to change their vocal ability, because they consider it to be unethical. Therefore the atrophy of the vocal cords of adult individuals is supplemented by an artificial device.. To reinforce the astuteness of the Ummo hearing, it should be noted that our Ummo visitors were able to identify, just by hearing, the different characters typed on their mechanical typewriters (E27, "he [an Oomomen] had noticed that each key has a different sound").

Consequently, it is possible that where we identify a single sound, Oomomen can identify two distinct sounds. This helps explain the many variations in "spelling" for the same word that is written phonetically. (See synthetic schemes on the deciphering process)

UMMO IDEOGRAMS

D 77: "The first language, DU-OI-OIYOO can be translated by Linking Language) uses ideograms in their graphic expression and phonemes groups related or connected representing concepts, values and concrete objects and even complex ideas ordered. This is a vehicle used to converse with routine (home language, technical, macrosocial popularized)."

The first-level of written language is ideographic. This is the only common point with the writing of the Asian and Chinese languages. It is reasonable to imagine that the coding of these ideograms was done using conventional correspondence that was probably conceptual. This is probable because without these details of correspondence between an ideogram and its conventional value, it would be impossible for anyone to decode the written language.

110

D731 T13-44/72	Background	Original ideograms	Ummo word
S731-s1	Each B.B sends universal, collective feelings, gregarious induction, and moral models, etc. to every OEMMII. YES, moral laws were "written" before the prophet sent by WOA on every planet with intelligent beings. The WAAM has an infinite mass divided in equal part of MATTER and ANTIMATTER. The photon speed inside is infinite.		BUAWEE BIAEI WAAM UU
S731-s2	Death of OEMII coincides with the disintegration of the OEMBUUAAW (the Kr atoms return to their quantum behavior) BI = BAA IYODUHU (Union factor between BB and the chromosomes		OEMBUAWE BAAIODUU
S731-s3	The brain receives instructions from the Soul		BUAWAA WAAM U
S731-s5	UIW is an ancient time unit used in OYAAUUMMO it is around 185.5 terrestrial seconds.		UIW
S731-s6	The 4th frontier effect "leeiyo WAAM"		LEEIYO
S731-f16-01	Divine wisdom		WOA
S731-f18	Signature: AOXIIBOO 3 (son of) IRAA 6		

A LANGUAGE WRITTEN IN PHONETIC

The words of the Ummo language consist of a "series of phonemes" and "the words noted in this document are approximate graphic expressions of their real voice." Oomomen explicitly indicate that their phonemes can, of course, be written with various spellings: "In most cases our phonemes can be made by several graphic expressions." The Oomomen tell us that the words are written phonetically in Spanish with various spellings.

D357-2:' The number of letters written means we stretch these phonetic sounds"

In addition there is therefore a repeat rule where the word doubles or sometimes triples the same letter to indicate long tones. This strictly phonetic writing can give more opportunities of different "spellings". In addition to the various possible spellings, the old texts were dictated to typists who introduced multiple deformities, such as the confusion of sounds that were close, typos, etc. The phonetics of the words was controlled, but initially there was no standard "spelling". The set of phonetic writing of a word along with its deformities is the set of what I call the "distortion spelling" of any original spoken Ummo word. Recent documents seem to follow a more rigorous standard "spelling", which coincides with the phonetics of the language used.

INDICATIONS IN THE TEXTS

D357-2 (Spanish): "Collective Soul or BUAUe BIAEII (the" e "is pronounced like a synthesis of A and E)

NR18 (French): 'OMGEEYIE (pronounced" omghééyié ")"

D21 [Spanish]: UM-MO [the "U" very closed and guttural, the M could be interpreted as a B]

D32 [Spanish]: It is the same for DAYS and YEARS. It should remove any doubt about it. In UMMO we use the phoneme XI or SI [it is difficult to find the appropriate letters] which means CYCLE, SPIN or REVOLUTION with a dual acceptance. That is to say that this is what you call a homophone word. With the word "XI" or "CSI" we express both UMMO rotating on its axis [one day] that such a wheel.

D41-3 [Spanish]: "Keep in mind that the period of rotation on its axis of our planet is a UMMO XII [English read: see]"

Sent from Paris D70—dictated by XOODOU-7: The words marked with [*] are in French in the original text and several elements prove the origin of a French typist [Rivera instead of Ribera, for example]. «...

D 69-3 [Spanish]: " . . . the phoneme XOOGU [the G is pronounced like an aspirated H] applies to the entire system."

D 58-4 [Spanish]:" . . . and we call BAAYIODUU [the Y is almost silent and D can be taken as a very soft Z]"

PHONETIC ALPHABET TABLE

I will use Spanish phonetics for reference.

Spanish Phonem possible	Spanish Phonem
A or E	A
M or N or V or B	B
D or T or Z or S	D
E	E
EE	EE
G	G
I	I
ll or Y	ll
K	K
L	L
M or N or V or B	M
M or N or V or B	N
O	O
R or G	R
S or Z or X	S
SZ or SD or ST	SD

Spanish Phonem possible	Spanish Phonem
D or T or Z or S	T
U	U
UU	UU
W	W
Y or ll	Y
X or GS or CS or KS	G + S ou K + S
D or T or Z or S	DS
WA or UA	W+A ou U+A

SOME EXAMPLES OF ENGLISH PHONETICS

D27 1966 ESP: YU 1, hija de AIM 368

D28 19/03/1966 ESP/ANG: YOO 1 daughter of AIM 368

NR18: [french: Oummain]. . . We use them interchangeably and in order of preference ooman, oomoman, oomoan in our correspondence with your English brothers.

A FUNCTIONALIST PHILOSOPHY

There is no ambiguity about the fact that the Ummo language is underpinned by a functionalist philosophy:

"Our thought has always been guided by a policy that has an analogy with the pragmatism of John Dewey Philosopher on Earth. We recognize a wide acquaintance with the highest level is occupied by the feature."

In Western languages the feature is mainly described by the verbs. Thus, if we express the main functionality of a wheel we say it "runs" for a vehicle he "moves".

To describe a functional concept of "wheel", we will have to express the functionality in a conceptual way. In other words, the concept of "spin". We can say that "turn" is a concept of "rotation", in other words, a more general way or "a period of a cycle."

To describe a vehicle as an object functionally, we could say that it is "an object which moves things" or it is "an object that produces a shift."

On the other hand, if we want to describe a vehicle as a functional concept, not as an object, then we would say that a vehicle is "that which is related to the concept of generating a movement." In this case we describe "vehicle" in an indirect way through its main functional concept.

As another example, if we want to describe the value 1 from a functional point of view, we may say functionally that it is to "identify a first occurrence."

The Ummo documents have thus far shown us that Ummo "words" were written phonetically as functional expressions, and they designate "concepts, values and concrete objects and even ordered complex ideas." In other words, Ummo "words" can describe anything.

*

Learn the Ummo Language!

Want to learn to speak the Ummo language?

Want to learn to count in Ummo language?

Here we will show you how to decipher the Ummo language yourself. Thus, you can read, write or count as the Oomomen do.

The method presented is also the one that helped me to decipher the Ummo language. The method is very complete and can be learned by anyone. After a detailed learning phase, the method is well understood and assimilated, and then its use is very simple. . .

CORE WORD STRUCTURE

Ref. document 104-1: "our texts are highly synthetic without morphology spelling that is familiar to you, which makes them very complicated decoding without prior addition of verb forms, adjective, etc."

Because the Ummo language has neither verbs, nor articles, nor pronouns, etc. it is almost impossible to translate directly. There is an enormous intellectual obstacle for those who wish to translate it. This obstacle is directly related, as we have seen, to the functional nature of the "words", and to the fact that they have been written phonetically with various "spellings". In addition, due to imprecision of expressions that are also strongly conceptual, we must add some words, as we will discuss later. Also it must be noted that we only know the culture of our visitors through these documents, so trying to understand their kind of thinking is a unique challenge that one has to face in making a semantic analysis of the Ummo language.

The Ummo "words" are NOT words in the usual sense. A single Ummo conceptualization as expressed in our language is the equivalent of a complete sentence. This sentence is made with small "bricks" that are the basic primary concepts. It would therefore be more accurate to call a "word" a "nested sequence of elementary phonetic concepts".

The "words" are written phonetically and each phoneme corresponds to a simple concept or primary idea. Each primary concept is embedded with the primary concept that follows it. Nesting is hierarchical: from the major primary concept on left to the primary minor the concept on right. It is mainly the nesting which gives the sense of the word. However, as in all languages, context is mandatory to complete understanding.

So, I experimented with a decoding and translating method of Ummo "words". This method was initially intended to clarify the meaning of the words in order to improve the understanding of the detailed documents. Except in some cases of uncertainty (which we will discuss later) this method determines the phoneme-matching concept actually corresponding to the context and spelling. The general testing method provides the spellings of long sounds and ambiguous sounds, and checks on whether or not the result is consistent with the context. Decoding is performed and the final translation of the word is identified for each group of concepts. After analyzing the words in this way, I found that:

. The words are phonetic, not "spelling"

. The words result from the nesting of primary phonetic concepts

It then followed that:

. A Ummo word is like a number. It must be read from LEFT to RIGHT, from the MAJOR concept to the nested MINOR concept. But for the analysis we must decode the word in the opposite side: from RIGHT to LEFT.

. In EACH NESTED PAIR A MEANING EMERGES which is related to the upper nested phoneme-concept.

. Consequently, only certain spellings for a phoneme are meaningful and not all spellings

THE PRIMARY PHONETIC CONCEPTS

The phonetic words are composed of 17 possible primary phonemes which follow the principle of "atomicity" proposed by Bertrand Russell and detected by Jean Pollion. Each primary phoneme combines a concept which is deduced and also functional. My research has led me to redefine the majority of the primary concepts of phonemes that were initially proposed. My method of determination is empirical, and is repeatedly checked through trial and error. I began by analyzing a word that seemed simple and clear, as taken from our understanding of the Ummo documents. Then I tested the identified concept of words that were more and more difficult.

Because of the fact that the written language was done phonetically, the primary concepts are also strictly phonetic. They associate a phoneme, not a letter, with a concept.

Spanish primary concept	General functional concept meaning	Some applications of the concept	Remarks
A		a) moving b) movement c) computable (value displacement) d) Process	Word type: OA moved entity Concept referring to cosmology (angular displacements, displacement resonance, etc.).
AA	Dynamic	a) Dynamic b) Moving c) Flow	ayuubaa d69-3: (ayubaa is a word equivalent to "network" or "structure" in dynamic binding). —moving in a topology are dynamically linked have dynamic interconnections

Spanish primary concept	General functional concept meaning	Some applications of the concept	Remarks
B		a) Interconnection b) Network node	IBOO D81: BETWEEN TWO IBOO (NODES OR CENTERS) —IDENTIFIES THE INTERCONNECTION WITH A MATERIALITY. —INTERCONNECTION MATERIALIZED
D	Form	a) Forms b) Appearance c) Demonstration d) Shape	
E	Concept	a) Concept b) Non-dimensional mental represen-tation (set of related mental pictures) c) Perception	
EE	Model	a) Model b) Design pattern	
G	Structure	a) Organization b) Structure organized	
I		a) Identification b) Identifies c) Uniqueness	The uniqueness of the identifica-tion is implied, other-wise there is no identifica-tion.
II	Limit	a) Limit b) Delineation c) Boundary d) Membrane	
K	Distance	a) Physical distance in 3D cosmos	Example of abstract distance "cultural distance"
L	Change	a) Change status	
M	Joint	a) Join b) Sum c) Adding	
MM		a) Join inseparable b) Sum c) Adding d) Junction, join permanent	The spellings "M" and "MM" can be phonetically distinct and meaningful.

Spanish primary concept	General functional concept meaning	Some applications of the concept	Remarks
N	Flow	a) Flow b) Transfer	
O	Entity	a) Dimensional entity	D41: Applies to that which is dimensional. (with time and space characteristic)
OO	Matter	a) Matter	
R		a) Surapposition	
S	Cyclicity	a) Cycle b) Alternating c) Induction d) Ripple e) Wave f) Rotation g) Repeating,	
T	Oriented Direction	a) Oriented sense of time	
U		a) Dependence b) Submission c) Influence	Injective relation
UU		a) Dynamic dependency	(U) dependence "a" (U) dependence = The dependence has a dependency = Dependency dependence = Dynamic dependence (Relates to the force fields, a parent-child relationship, food addiction, etc.) DUU: "attraction"
W		a) Generation b) Generating c) Emergence d) Create d) Generating e) Produce	
Y	Spatiality	a) Spatiality b) Spatial c) Topology d) Space	Scheduling in space, surfaces, volumes.

THE NESTING AND THE BASIC RELATIONSHIP "GET"

Each pair of concepts is nested from LEFT to RIGHT, from the MAJOR concept to the nested MINOR concept. But for the analysis we must decode the word starting with the opposite side: from RIGHT to LEFT. In order to achieve a translation, the relationship between two nested concepts can be INTERPRETED by using the word "get". Of course, the "get" is an artifice of translation that only serves to prioritize the concepts properly. This translation of the relationship with "get" makes it possible to rigorously schedule and follow the nested concepts. This device was used in the first stage of the decoding of the Ummo concepts. Using it resulted in a basic literal synthesis. The second step was to translate the literal synthesis into the target language.

Regarding Cn (or the primary phonetic concept) the basic relationship is the following recursive sequence:

$$Cn + 2 = Cn + 1 \text{ "a" } Cn$$

THE GRAPH OF THE RELATIONSHIP OF PRIMARY CONCEPTS

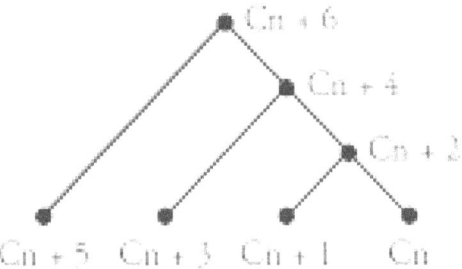

Graph of the fundamental structure of the primary phonetic concept

THE NEGATION

There is no denial value, but rather the assertion of the absence which is expressed by the logical value, "WRONG-NOTHING VALUE". This is expressed by the word or segment EEDOO translated as the "model or conceptualization of a material form", that is to say something that was only a mental representation.

Examples:

KEOYEEOO

IIAS IBOZOO UU AIOOYEEDOO

YAEYUEYEDOO

THE MULTIPLICITY

The doubling of phoneme sequences expresses the multiplicity, that is to say several times something that is functionally equivalent, but not necessarily of the same nature. There is no plural marker itself.

Some examples:

WOI WOI

WAAM WAAM

CONTRACTIONS SPELLING

When linking two words together into one, there may be a contraction spell. For example, the word UMMOEMMII is a contraction spell of "UMMO"—"OEMMII", and is simply the result of phonetics.

THE REFLEXIVITY

In Spanish a letter may often be associated with a concept. Oomomen also tell us they have noticed the long Spanish sounds by doubling letters.

Conceptual reflexivity is expressed by the doubling of primary phonemes. It generally reflects the "dynamic". Some examples (see details in "The primary phonetic concepts"):

a. (U) dependence "get" (U) dependence

—The dependence has a dependency

—Dependence of a dependency

– Dynamic Dependence

This dependence can be described as concrete or abstract.

b. (I) identification of "get" (I) identification

—The identification has an identification

—Identifying a "side" has an identification on the other "side"

—Limit, frontier

c. (E) concept "get" (E) concept

—The concept has a concept

—Concept of concept

—Model or Mental pattern

d. (O) entity "get" (O) entity

(see The primary concepts "O—entity" et "OO—matter")

—The entity has an entity

—entity with mass

—Material, matter

THE OPERATOR "AND"

When the concept "(M) join" is related to two concepts or identified segments, nested and minors (notice that, it is not necessarily the basic concepts), the (M) "break" the nested hierarchical. The first two concepts are reflected at the same level as they are "joints" on the same plane. In this case the concept "(M) join" acts as an "AND". The result of the join is the superior concept (C4 in this example):

To give an example OMGEEYIE is the concept of "couple" represented as an "object": (O) is the entity formed by the junction of a GEE and a YIE.

(C4) "get" [(M) "get" [(C1) "get" (C2)]

= (C4) "get" [the joint get [(C1) "get" (C2)]

= (C4) "get" [(C1) "AND" (C2)]

Notice that the concept (M) "join" is inclusive, whereas the concept (U) "dependency" relationship is a non-inclusive.

Spellings (M) and (MM) can be phonetically distinct and orthographically meaningful, for example:

- "OEM-MII" which is pronounced in english "oem"—"mee". Here the first (M) ends a logical segment and follows two concepts.

- and "OEMII" which is pronounced in English "oemee". Here we have a simple (M), and (II) which is linked to the concept of "body" by the only nesting hierarchy.

- Unlike OEM and OEMMII, the word "UMMO" must to be read in English "oom"—"mo", we only have one concept before (MM). Therefore, it stands to reason in this case that the spelling (MM) may be meaningful and describe the "dynamics", which is a true long "m".

*

Decode and translate the Ummo words

In the original Ummo language words are written with ideograms.

From the word phonetically written in Latin letters, I experimented with a method of decoding and translating the Ummo words. This method was initially intended to clarify the meaning of words, to improve the detailed understanding of the documents.

The principle of this method is to analyze the word into two steps::

- . A decoding step
- . A translation step

First we need to decompose the concepts (DECODING) from the words which are phonetically written in Latin letters. Then we must seek to understand what they refer to (TRANSLATION) in the context of Ummo culture. Some spellings are significant and others not. If we do this, we will then find the sequence of letters which corresponds to the real meaning of the Ummo word.

If we work in this way, the final completed translation and the initial definition will need to agree with each other. The context will be the tool which allows us to translate the generic concepts of the functionalist Ummo culture into object-words which follow the Earth standard. (Or at least the Western standard.)

DECODING

The goal is to express the concept resulting from the spelling of the word while also knowing that words (as we have seen) are the result of the relationship between phonemes and concepts:

- . The words are the result of LINKED PAIRS of NESTED CONCEPTS (primary concept or established, or proposal concept).

- . A Ummo word is like a number. It must be read from LEFT to RIGHT, from the MAJOR concept to the nested MINOR concept. But for the analysis we must decode the word on the opposite side: from RIGHT to LEFT.

- . In EACH NESTED PAIR THERE EMERGES A MEANING which is related to the upper nested phoneme-concept.

- . For TRANSLATING PURPOSES, the relationship between two phonemes-concepts can be translated in English as "get". This translation of the relationship with "get" allows the arranging of the decoding in a way that completes the translation of the word.

THE TRANSLATION

The transition from the Ummo functional concept to the terrestrial word object is a difficulty which must be conquered in order to complete the semantic analysis of the Ummo language.

The final translation of the text and the Ummo definition must coincide. However, the context itself will allow us to translate the generic concepts of the functionalist Ummo culture into object-words which following the Earth standard . . . if the word expresses something intelligible for us within that context.

*

THE GENERAL METHOD OF SEMANTIC ANALYSIS

THE SEMANTIC ANALYSIS OF WORDS: PROBLEMATIC

So in essence we have words that combine a phoneme with a basic concept, inside an overall phonetic writing. This phonetic script can be written with various spellings and it can give us a set of "distortions or variations in spelling" for any given word. The question therefore is whether this language has...

- 1 phonetic word for 1 concept?

- 1 phonetic word for 2 different concepts?

- 1 phonetic word, 1 form that is orthographically meaningful, and 2 different objects which are conceptually equivalent?

- 1 phonetic word with a meaningful orthographic form that is indeterminable?

To respond to the points of the problematic, I have designed and tested a general method which begins to test the spelling of long sounds and ambiguous sounds, and to check if the result is consistent with the context or not. The method then performs the decoding and the final translation of the identified word for each group of concepts.

IDENTIFYING CONCEPTS

With several cases:

a. 1 phonetic word for 1 concept

b. 1 phonetic word for 2 different objects, but conceptually equivalent

c. 1 phonetic word for 2 different concepts, depending on the verbal context

IDENTIFICATION SPELLINGS

With several cases for a phonetic word:

a. 1 significant orthographic form, 1 concept or object

b. 1 significant orthographic form, 2 different objects conceptually equivalent

c. Significant orthographic form(s) which is (are) indeterminable

IDENTIFICATION OF SOUNDS

With several cases:

a) Identification of ambiguous sounds

. with the phonetic table

b) Identification of long sounds, by cross-comparisons with the words

. with another clearly distinct concept

.   in texts with separate editorial dates

.   in texts with different authors

DECODING AND WORD TRANSLATION

. Decoding to confirm the significant spelling
. Translation to get the current terrestrial sense

THE IDENTIFICATION OF CONCEPTS IN UMMO TEXTS

For the Ummo words designating "concepts, values and concrete objects and even complex ideas ordered" we will first seek to identify the concepts in the texts that the phonetic word(s) could express.

We must therefore identify the concept associated with each word and constitute a subset of "distortion spelling" for each concept.

With several cases:

. 1 phonetic word for 1 concept
. 1 phonetic word for 2 different concepts

IS THERE 1 PHONETIC WORD FOR 1 CONCEPT?

Take the example of the concept of "female" pronounced "yea".

Here is the table of phonetics occurrences with distinct spellings.

N°	Date	Language	Word	Extract
D41-6	1966	ESP	IIE	the IIE (WIFE)
D 541	10/04/1987	ESP	IIEE	IIEE I am a woman of Ummo (We pronounce IIEE to express our gender)
D41-11	1966	ESP	YIE	the YIE (woman)
D 1378	30/01/1988	ESP	YIEE	that YIEE (woman)
D176	27/02/1983	ESP— from Malaysia	YIES	your Yies (women)
D102-1	16/10/72	ESP	YIHIE	the percentage of YIHIE (women)
D174	15/01/1981	ESP	YIIE	hello to your distinguished YIIE (wife)
	02/04/1993	ESP	YIIEE	And also their YIIEE.

Is it reasonable to imagine that we have here a profusion of distinct concepts for "female"? Of course the answer is no.

It is more likely that most sounds, and especially the long sounds, were simply written in numerous ways by various people following the Spanish phonetics. In this simple case it is easy to identify 1 phonetic word for 1 concept.

IS THERE 1 WORD FOR 2 OBJECTS CONCEPTUALLY EQUIVALENT?

Is there 1 phonetic word for 2 different objects that are conceptually equivalent?

We have the answer directly in the document below:

Letter D357-2 (ESP): "The confusion you may see comes from what we call BB (BUAUEE BIAEEII) not only the collective Soul of UMMO or Earth, but also the cosmic plan (that is to say the multicosmos) that contains all BB [the WAAM-UU] of all different social networks that populate our tétra-dimensional cosmos."

It is therefore possible to have 1 phonetic word for 2 different objects that are conceptually equivalent.

IS THERE 1 PHONETIC WORD FOR 2 DIFFERENT CONCEPTS?

We have the answer directly in the document below:

Letter D69-2 (ESP): "The phoneme OOLEA has a different meaning depending on the verbal context in which it is integrated. The most correct meaning when applied to a technical field is as follows: change, change from one physical environment to another.

In scientific language it means: increase or decrease the value of an angle in another infinitesimal angle. This would, in the case that we are studying, the most faithful version of the phonetic root."

It is therefore possible to have 1 phonetic word for 2 different concepts. The word would then be a homonym whose meaning depended on the context in which the word was integrated.

CONCLUSION ON IDENTIFICATION OF CONCEPTS

The identification of concepts is the first and most important step to decode Ummo words.

We therefore have:

a. 1 phonetic word for 1 concept

b. 1 phonetic word for 2 different objects that are conceptually equivalent

c. 1 phonetic word for 2 different concepts, whose meaning depends on the verbal context

Regarding point c), we cannot generally characterize the homo-nym with our present state of our knowledge. This shows the limits that we have in decoding words, due to the fact that the meaning of some of them is linked to the context of the "sentence".

Only preliminary explanations from the authors can remove the uncertainty. The decoding of words would be optimal if we could clearly identify 1 phonetic word for 1 concept.

THE IDENTIFICATION OF SPELLING

In order to decoding a Ummo word-even in the simplest cases where there is 1 phonetic word for 1 concept—we must determine the phoneme associated with the significant spelling. For each orthographic form associated with one concept, we have various possibilities of letter sequences, with the following simplified gene-ral cases:

a) The number of letters in the word corresponds exactly to the number of Spanish phonemes (letters Nb = Nb phonemes). This is the simplest case where each letter written in Spanish pro-duced a phoneme-concept. We can then assume there is not long sound and simply do the decoding.

b) The number of letters in the word is greater than the num-ber of Spanish phonemes (letters Nb> Nb phonemes). This is the most frequent case. Most likely this scenario indicates the presence of one or more long sounds.

c) The number of letters in the word is less than the number of Spanish phonemes (letters Nb <Nb phonemes). This is the rarest cases. Most likely it indicates a typing error.

These three general cases are simplified and trivial, because phonetic Spanish is a bit more complicated.

Some sounds can be ambiguous. For example, in Spanish the isolated letter "u" gives us the English sound "oo". However the association of the letters "WA" or "AU" will give both the Spanish sound "U"—"A", which is to say "wa" in English. We may use two tools to overcome this difficulty:

. The phonetic table will allow us to identify ambiguities of this nature.

. The decoding and translation method will determine the phoneme-concept which actually matches with the context.

We can see this in the case of the word BUAWA (and it is also characteristic of the word YIE.) It is clear that there are not a multi-tude of different concepts for the idea of "Soul", which is expressed by the word BUAWA. Here we are in the case where there is 1 pho-netic word for 1 concept. We must determine the phoneme asso-ciated with significant spelling among its many different spellings. Here is the table of distinct spellings:

Réf. D	Date	Language	Word	Extract
NR-20	17/01/2004	FR	BOUAWA	. . . we outsource partially that to entities transcendent as are the individual soul (BOUAWA) the collective psyche (BOUAWEE BIAEII) and God (WOA).
D43	1966	ESP	BUAAWA	. . . and the BUAAWA (SOUL) creates only IDEAS and directs our OEMII (body)
D 792-1	janvier 1988	ESP	BUAAWA	OEMMIIWOA has classics BAAYIODUU networks incorporated by atoms of Krypton which are in connection with his brain and his BUAAWAA with the BUAWWEE BIAEII.
D105-1	12/07/72	ESP	BUAUAA	we know that the "soul" that we call BUAUAA is dimensionless and therefore the TIME has no meaning for her.
D105-1	12/07/72	ESP	BUAUAAA	phoneme BUAUAAA
D 1751	14/01/1991	ESP	BUAUUAA	Maybe if they had studied the soul (OOA EESEE BUAUUAA) of this dictator, they would have been much more judgment elements.

Réf. D	Date	Language	Word	Extract
D 791	27/12/1987	ESP	BUAWA	Sometimes the pressure of the second flow (from: Outside the world and internal environment) or that of the fourth flow of subconscious origin) are so intense that they saturate the threshold for action of BUAWA through the quantum structure krypton, and free a decision is impossible.
D 21	mai-66	ESP	BUAWAA	BUAWAA (SOUL).
D520	22/11/1988	ESP	BUAWAAA	And soft, in a long time, the day where your BUAWAAA melts into a tight embrace with that of your beloved son.
D357-2	12/03/1987	ESP	BUAWUA	B. (BUAWUA) IS A CELL CLOSED ON ITSELF FROM BILLIONS CELLS IN THIS WAAM.
D357-2	12/03/1987	ESP	Buawuaa	The Buawuaa (SOUL) is not capable of processing data, to think, to develop information, but only to maintain, to record.
D357-2	12/03/1987	ESP	BUAWUUA	BUAWUUA BIAEEI or collective psychic brain
D 792-1	janvier 1988	ESP	BUAWWA	his BUAWWA
D33-3	1966	ESP	BUAWWAA	concepts dimensionless as BUAWWAA or BUAWE BIAEI (SOUL AND COLLECTIVE SPIRIT)
D105-2	12/07/72	ESP	BUUAUA	description of BUUAUA
D105-2	12/07/72	ESP	BUUAUAA	BUUAUAA (individual mind)
D 541	10/04/1987	ESP	BUUAUUA	BUUAUUA as this cosmic cell is called shape the whole conduct of man freely and at once,
D 541	10/04/1987	ESP	BUUAUUA	his BUUAUUA
D 1751	14/01/1991	ESP	BUUAUUAA	He wants to reassure his mind (BUUAUUAA)
D 33-1	18/03/1966	ESP	BUUAWA	AIOOYA AMMIE BUUAWA: SOUL EXISTS.
D 31	16/03/1966	ESP	BUUAWAA	In the early stages of humanity these men, who some years before, WOA (God or generator) gave the influence of an unidimensional being, as is the BUUAWAA (soul), cease to be simple anthropoid animals.
D41-15	1966	ESP	BUUAWAAA	our BUUAWAAA

Réf. D	Date	Language	Word	Extract
D41-15	1966	ESP	BUUAWEA	BUUAWEA (SOUL).
D357-2	12/03/1987	ESP	BUUAWUA	is WAAM stores the entire constellation of BUUAWUA (souls or spirits) of all human beings of our universe.
D357-2	12/03/1987	ESP	BUUAWUAA	BUUAWUAA BIAEII (COLLECTIVE PSYCHE)

THE IDENTIFICATION OF LONG SOUNDS

The long sounds are normally written with a doubling phoneme. With the transposition errors in Spanish, the long sounds are difficult to identify. As we have seen, when the number of letters in the word is greater than the number of Spanish phonemes (letters Nb> Nb phonemes) we can deduce the presence of one or more long sounds. However, when there is at least one of the orthographic forms of the word which is denoted by a double letter, we can logically question if there really is a long sound. Again, we have many tools to overcome this difficulty.

We can do cross-comparisons:

. with words whose concept are clearly distinct
. with words in texts on separate editorial dates
. with words in texts by different authors

The identification of long sounds is obviously more probable if the spelling of the word is repeated in similar words, at different times of writing and by different authors.

. If this is done then the decoding method and translation will determine the phoneme-concept corresponding to the actual context, since we know that these long sounds mark the "sequentially repetitive nature of the concept."

In some cases, we can detect the difference between a short or a long sound, since there is a real phonetic difference between them. This phonetic difference corresponds to a difference in the concept meaning, so we also must find two groups of words—a group with 1 letter for the short sound, and a group with 2 letters for the long sound group.

Below is an example of this using "OEMMII" and "OEMII". These are 2 different words with 2 different meanings, but they are not so easy to identify.

Réf. D	Date	Language	Word	Extract
D41-15	1966	ESP	OEEMII	The OEEMII (BODY-SOMA).
D381	??/02/1988	ESP	OEEMMI	. . . not stood up to defend abused OEEMMI.
D 89	01/03/1969	ESP	OEMI	other OEMI on Earth.
D33-2	1966	ESP	OEMII	PHYSICAL BODY of MAN
NR-13	14/04/2003	FR	OEMII	understanding the WAAM WAAM requires the full biopsychosocial understanding of the OEMII (human body in its material and psychic inseparable aspects).
D33-2	1966	ESP	OEMIII	this marvelous organism that WOA has created, which is the OEMIII
D 731	20/03/1987	ESP	OEMIIS	for any OEMIIS.
D104-2	19/02/73	ESP	OEMIS	But the mentality of these oemis (MEN) makes such a utopian hope.
D357-1	12/03/1987	ESP	OEMMI	the model of the WAAM-WAAM (cosmos multiplanar) which contains intelligent beings OEMMI.
D 1751	14/01/1991	ESP	OEMMIE	It is only when the mass of the human cortex (OEMMIE) has evolved that its social network will be free from this slavery.
D 33-1	18/03/1966	ESP	OEMMII	OEMMII (HUMAN BODY) IN THE WAAM (COSMOS).
D 731	20/03/1987	ESP	OEMMIII	they are not able to move towards OEMMIII (human).
D357-2	12/03/1987	ESP	OEMMIIS	The OEMMIIS on Earth
D 731	20/03/1987	ESP	OEMMIS	Two OEMMIS from different global social networks cannot procreate
D 1378	30/01/1988	ESP	OENMMII	my brothers decided to form very small groups of OENMMII of different nations of the Earth,
NR20	17/01/04	ESP	OEMII	In the conformation process of our BOUAWA (Soul) after the disappearance of our OEMII (body).

By their distinct phonetic "oémi" and "oém"—"mi", we identify two distinct concepts, the concept of phonetic words "oemi":

D33-2	1966	ESP	OEMII	PHYSICAL BODY of MAN

The concept of phonetic words "oem"—"mi":

D 731	20/03/1987	ESP	OEMMIII	they are not able to move towards OEMMIII (human).

Considering all of this we might say that:

OEMII (the PHYSICAL BODY OF MAN) + "Soul" (BUAWA) = OEMMII (human).

So regarding the phonetics words "oemi" et "oem"—"mi" we have thereby identified:

a) 2 concepts: "human beings" and "human body caught in its material aspects"

b) 2 subsets spelling OEMMII and OEMII

(See detailed semantic analysis OEMMII and OEMII in UMMO Dictionnary)

THE EXPECTED RESULTS

The final result possible for a phonetic word is as follows:

a) 1 orthographic form that means 1 concept or object
b) 1 orthographic form that means 2 different objects that are conceptually equivalent

c) The meaning of the spelling is unclear:
—It could be 1 phonetic word for 2 different concepts, depending on the verbal context

—There are not enough elements to conclude the analysis

—There are several concepts that are not explicit in the texts

SYNTHETIC SCHEMES FOR THE SEMANTIC ANALYSIS OF WORDS

General diagram of the processing stages of the words

Ummo language to spanish

Ummo words in spanish phonetic

TRANSCRIPTION of primary concepts

TRANSLATION in terrestrial words

Detailed diagram of words transformations

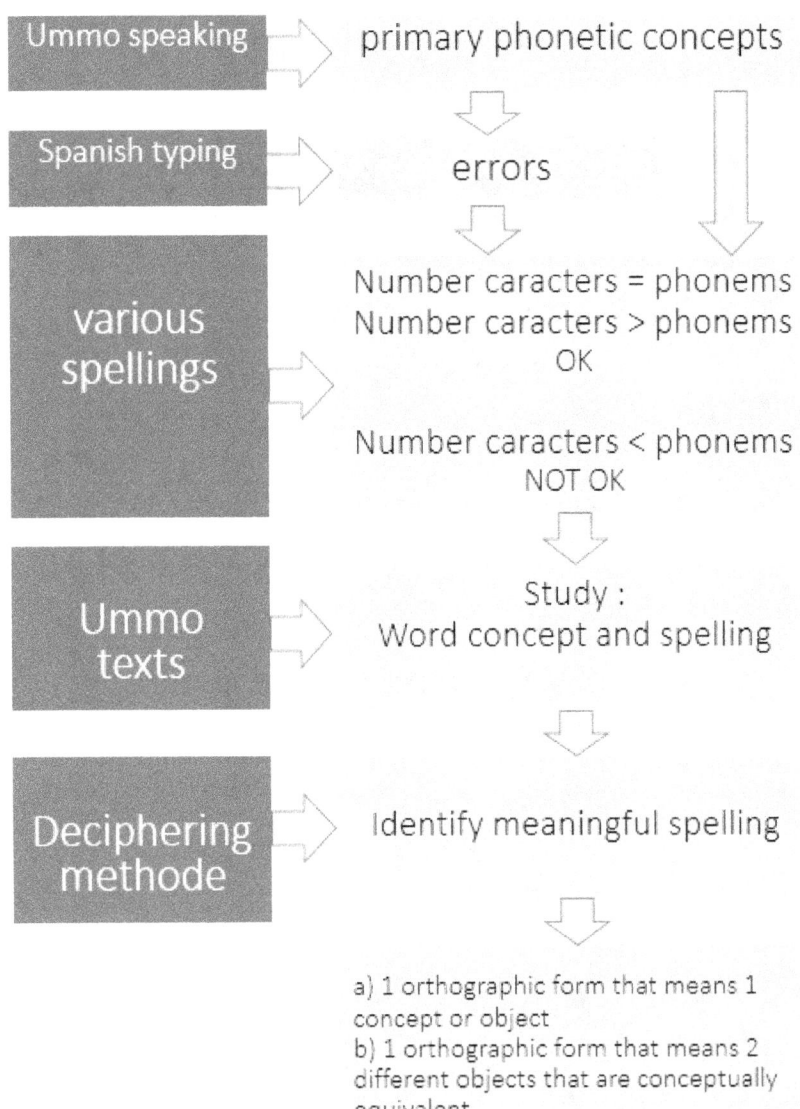

SOME EXAMPLES OF WORDS DECODING

Here are two sets of examples of words that are conceptually nested in several levels.

The first example is a standard case. The word "csi" describes the simple concept of "rotation" which is explicitly applied to describe the rotation of a wheel, or the rotation of a star on itself.

The second example involves the words "oo-a" UUA, "w-oo-a" WUUA and "oo-w-oo-a" UWUUA whose conceptual nesting levels are extremely strong, and thus show the extraordinary consistency of this conceptual language. This series of nested words describes concepts of "law", "mathematics" and "logic".

DECODING THE PHONETIC WORD "CSI"

Let us consider a simple example and teaching instrument that requires no special knowledge to be understood. The phonetic word "csi" describes the simple concept of "rotation" that is very explicitly used to describe the rotation of a wheel, or the rotation of a star on itself.

We know that:

. The words are the result of LINKED PAIRS of NESTED CONCEPTS (primary concept or established, or proposal concept).

. A Ummo word is like a number. It must be read from LEFT to RIGHT, from the MAJOR concept to the nested MINOR concept. But for the analysis we must decode the word on the opposite side: from RIGHT to LEFT.

. In EACH NESTED PAIR THERE EMERGES A MEANING which is related to the upper nested phoneme-concept.

For TRANSLATING, the relationship between two phonemes-concepts can be translated in English as "get". This translation of the relationship with "get" allows the ordering of the decoding in order to complete a translation of the word.

THE CONTEXT OF THE PHONETIC WORD "CSI"

Réf. D	Date	Language	Word	Extract
D 32	18/03/1966	ESP	CSI	On UMMO we divide the CSI (DAY) in UIW (1 UIW = 3,092 minutes) and there are no divisions corresponding to the hours.
D 32	18/03/1966	ESP	XI	a "XI" = 600.0117 (about six hundred UIW)
D 32	18/03/1966	ESP	XI	a "XI" (day of UMMO) = 1,855.2 minutes, that is to say a few 30.92 hours on land.
D 32	18/03/1966	ESP	XI	On UMMO we use the phoneme XI or SI (it is difficult to find the appropriate letters) meaning SPIN CYCLE or REVOLUTION has a dual acceptance. That is to say that it is what you call a word HOMOPHONE. With the word "XI" or "CSI" we express both the UMMO rotation of its axis (A DAY) that such a WHEEL.
D 977	20/06/1967	ESP	XII	Since many XEE (We call a XEE = 60 or XII periods of rotation of the planet).
D 977	20/06/1967	ESP	XII	vehicles whose propulsion and control equipment was each xii, more compact,
D41-12	1966	ESP	Xii	URAA (chronicles of this time) tell of the frightful scenes of Xii (days)
D41-3	1966	ESP	XII	UMMO our planet on its axis is a XII (read SII) equal to UIW 600, equivalent to 30.92 hours.
D41-3	1966	ESP	xii	DAY or our "xii" is divided by 600 we UIW
D41-7	1966	ESP	Xii	the first Xii (DAYS)
D41-9	1966	ESP	XII	the proceedings of XII (UMMO day)
D41-9	1966	ESP	XII	the XII (DAY) the divine UUAA (RELIGIOUS LAWS)
D47-1	1966	ESP	XII	Citizens were advised regulating the UAXOO in is forcing itself, without any constraint, to practice UIW 30 in all Xii (UMMO DAYS)
D57-1	23/01/1967	ESP	XII	XII (xii is a rotation or "DAY" UMMO)
D57-1	23/01/1967	ESP	xii	FREQUENCY UNIT (In periodic phenomena): This is the "XII", whose translation has various meanings for this word is not only the DAY UMMO but also the rotation of a wheel or the cycle time per unit.
D57-4	20/03/1967	ESP	Xii	have allowed our brothers to survive for 240 XII (Xii an equivalent to a rotation of our planet, few 30.9 hours).
D68	04/07/1967	ESP	XII	Few XII (Day UMMO) after he received permission from the UMMOAELEWEE

Réf. D	Date	Language	Word	Extract
D74	??/03/1969	ESP	XII	Rotation on its axis: 30.92 hours (we measure UIW: UIW 600 = 30.92 hours). (This is equivalent to a XII—see note 1).
D74	18/03/1966	ESP	XII	(The phoneme XII is a homophone that expresses both the duration of the "day of UMMO" as a "cycle", a "revolution" a "rotation unit", etc..).
NR-18	19/12/2003	FR	XII	We measure time as you depending on the path apparent from our sun during a IOUMMA XII (day).

IDENTIFYING THE CONCEPTS OF "CSI"

D 32	18/03/1966	ESP	XI	On UMMO we use the phoneme XI or SI (it is difficult to find the appropriate letters) meaning SPIN CYCLE or REVOLUTION has a dual acceptance. That is to say that it is what you call a word HOMOPHONE. With the word "XI" or "CSI" we express both the UMMO rotation of its axis (A DAY) that such a WHEEL.

The phoneme "CSI" is the concept of "rotation".

IDENTIFICATION OF THE SPELLINGS OF "CSI"

NR-18	19/12/2003	FR	XII	We measure time as you depending on the apparent path from our sun IOUMMA during a XII (day).

For this concept we have the spelling which seems clearly identified to XII.

THE IDENTIFICATION OF AMBIGUOUS SOUNDS OF "CSI"

The sound "csi 'is potentially ambiguous.

See Tables phonemes and phonetic primary concepts

THE IDENTIFICATION OF LONG SOUNDS OF "CSI"

The sound "i" is long.

DECODING XII

XII = (G) structure "get" [(S) cyclicality "get" (II) limit]

Relations between phonemes	Decoding the relation between functional concepts	Literal synthesis
S "get" II	[(S) cyclicality "get" (II) limit]	cycle delimited
G "get" SII	(G) structure "get" [(S) cyclicality "get" (II) limit]	The structure has a cycle of determined limits.

For XII we have the decoding:

. The structure has a cycle of determined limits

THE GRAPH OF XII

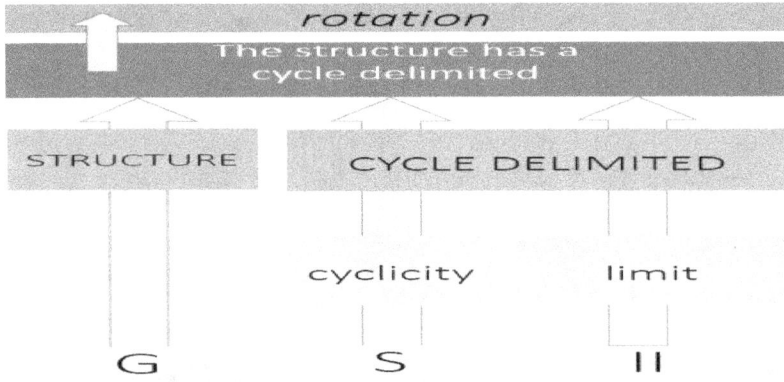

THE TRANSLATION OF XII

. The structure has a cycle of determined limits.

In other words, in that context: rotation

CALCULATING THE PROBABILITY OF XII

Note that if the word had been constructed by mere chance from the 17 primary concepts, the probability of having this combination would be:

1/17 3 = 1/4,913 = 0.0002035

CONCLUSION ON XII

The meaningful spelling of XII is transcribed as "The structure has a cycle of determined limits". As our Ummo friends suggest, this may indeed be easily translated into our terrestrial word: "rotation".

DECODING THE WORDS UUA, WUUA, UUWUUA

The second example deals with words including nesting conceptual levels have an extraordinary coherence. This series of words describes the concepts "function", "mathematic", "logic".

We have the following concepts:

a. UUA   = function

b. WUUA   = mathematic

c. UUWUUA = logic

We have the hierarchical nesting concepts as follows:

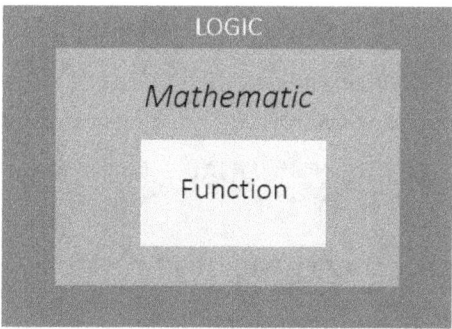

In a synthetic way we have:

a. UUA = function

—The dynamic dependency [has movement]

—The dynamic dependency has a [change-information-value-result]

—Inseparable with a [change-information-value-result]

—Hence this word must mean function

Note that if the word had been constructed by mere chance to use the 17 primary concepts, the probability of this combination is:

1/17 3 = 1/4,913 = 0.0002035

b. WUUA = W "get" [UUA] = mathematics

Decoding WUUA: "Generates UUA"

We can easily see that the word WUUA consists of: (W) generation "get" function

—Generates functions

—Hence this word must mean mathematics

Note that if this word had been constructed by mere chance to use the 17 primary concepts, the probability of this combination is:

1/174 = 1/83521 = 11,197 x 10-5

c. UUWUUA = UU "get" [W "get" [UUA]] = UU "get" [WUUA]

So we have a third word perfectly consistent with the two previous ones. Similarly, the transposition is:

UUWUUA = UU "get" [WUUA] = (UU) dynamic dependency "get" [mathematics]

= Dynamic dependency with mathematics

= Inseparable from mathematics

= Hence this word must mean logic

The final translation of the conceptual word UUWUUA as the Ummo people are using it in the context of their documents is "logic".

The hierarchical nesting concepts can be represented by the following scheme:

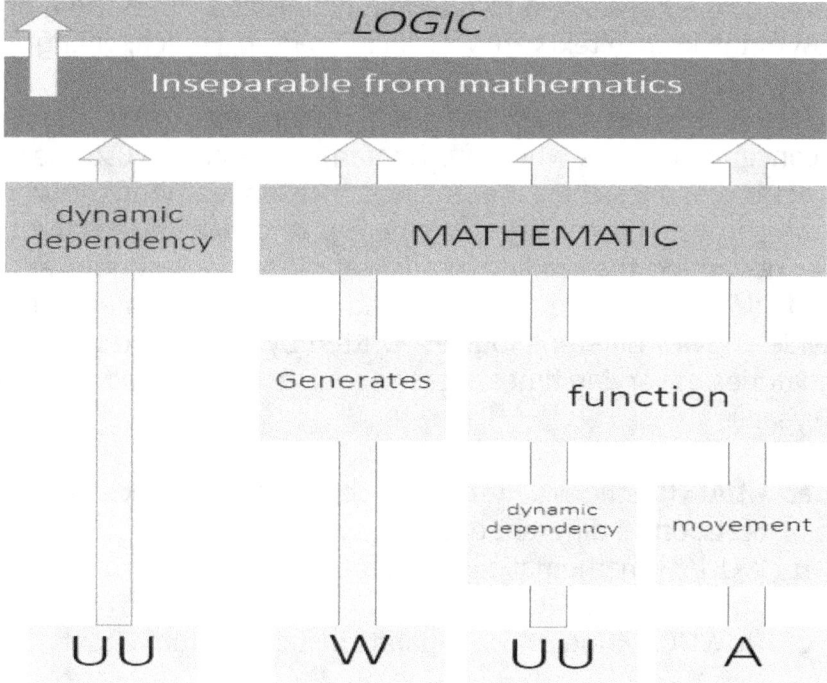

Note that if this word had been constructed by chance to use the 17 primary concepts, the probability of that combination would be:

$$1/17^6 = 1/24137569 = 41{,}429 \times 10^{-8}.$$

That means about 4 chances in 10 million. So if this language was indeed invented by a joker, he is one lucky guy and he should also play the lottery!

THE OVERALL CONSISTENCY OF WORDS

There is no reason to believe that the information given by the Oomomen about their language is misleading. We can decode both the sentences and the words by following the same general principle.

There are 7,503 entries in the database of known Ummo words, which come from about 250 documents and more than 1300 pages. The dissemination of these documents was conducted over a period of 40 years and to multiple countries around the globe. Most of the known texts are written in Spanish, French, or English.

Despite the difficulty presented in understanding the meaning of conceptual words when our human minds are accustomed to object-based ones, there is a striking consistency among the words in the corpus. This consistency emerges once we understand the rules governing the reading words. Let's look, for example, at the word UUA—which we find occurring in many documents, from a wide array of different sources, written by varied and unrelated signatories, at varying times over the course of nearly half a century (40 years):

a. UUA = functions (laws) => 108 separate occurrences, from 24 texts, on 16 different dates
b. WUUA = mathematic => 14 separate instances, from 9 texts, on 6 different date
c. UUWUUA = logic => 2 separate instances, from 2 texts

As we can see from the very simple example of these three words, the coherence is extremely strong. The probability of getting this result by mere chance is equivalent to the probability of having...

"A tornado assembles a plane by ripping through the public dump."

This consistency is neither coincidence nor chance, but rather a logical system.

SENTENCE STRUCTURE OF THE LANGUAGE "DU-OI-OIYOO"

THE CONTEXT

D77: ". . . That kind of language not uses "words": the proposals encode components of the sentence by agglutinating (subject, verb and predicate as you would say) in the form of coded proposal."

EXPLANATION FOR EXAMPLE

Here is a quote given to us by the Oomomen: "OEMMII OIAGAA GAEOAO AIOOIAO OEMII UIAA OEMII EABAYO UAMM. . . IIA"

It translates to: This man is noble. This man loves the Universe: That's enough.

We have the following nested series of linked concepts, or concatenation, which is signified by the spellings: "OEMMII OYAGAA GAEOAO AIOOYAA OEMII UIAA OEMII EABAYO WAAM. . . IIA"

Let's see what the decomposition means, word for word:

Word		Concept
OEMMII	(Humans)	"The body is associated with a Soul"
OYAGAA	(planet of the square signal)—Earth	"The entity that has an orbit and has a dynamic structure"
GAEOAO	(psycho-technical formula)	"Description of socio-professional skills"
AIOOYAA	(True-Positive)	"Entity identified in real-dimensional"
OEMII	(man)	"Physical Body of Man"
UIAA	(decision)	"subordination link"
OEMII	(man)	"Physical Body of Man"
EABAYO	(love)	"The thought is channeled on a set of things"
WAAM	(cosmos)	"Dynamic and generation simultaneous"
IIA	(limit moved)	limit moved . . . to the end.

{[Humans—Earth] {[True-Positive psychological formula] {[man—decision] {[man [love—cosmos] {[moved limit]}

THE DECODING OF THE SENTENCE

What one can fairly accurately transcribe is:

. Humans on Earth have a positive psychological formula.
. The man decides.

. The man loves the cosmos.
. End.

THE CONCEPTUAL GRAPH OF THE SENTENCE

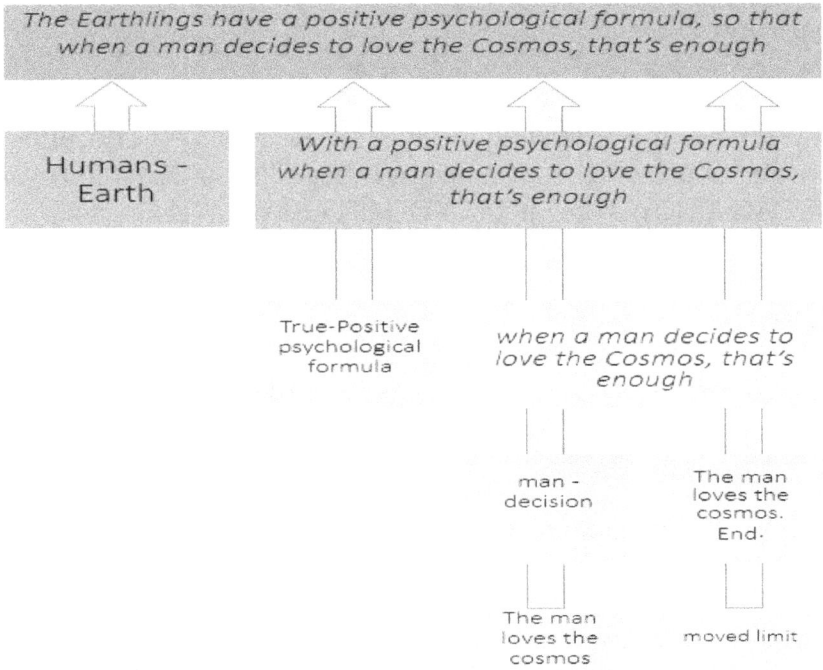

THE TRANSLATION OF THE SENTENCE

We can translate by: The Earthlings have a positive psychological formula, so that when a man decides to love the Cosmos, that's enough.

NOTES ON SENTENCE

It may be noted that in the initial Oomomen translation, our visitors have:

1. Truncated one word:
   "OEMMII OIAGAA" becomes simply "man"

2. Contracted two words:
   "GAEOAO AIOOIAO" becomes simply "noble"

3. Contracted two sequences:
   "OEMII UIAA OEMII EABAYO UAMM. . . IIA" becomes "This man loves the Universe: It's enough."

Note that our friends gave us a very approximate translation, without respecting the structure of their original "sentence".

PROBABILITY CALCULATION FOR THE SENTENCE

Note that if this sequence of words was built by chance with 17 primary concepts, the probability of this combination is:

$$1/17^{50} = 1 / (3.33 \times 10^{61}) = 1 / 33.30 \text{ billions of billions of billions of}$$ billions of billions of billions. I hope that this fact brings an end to the comments from the naysayers.

COMMUTABILITY IN SENTENCES

Words in a sentence are commutative. As an example let us consider the sentence, "space ship changing is tridimensional frame to get from one planet to another". We have seen that there are several possible constructions:

a. UEWA OEMM OAWOOLEA
b. OAWOOLEA UEWA OEMM
c. OAWOOLEA oemm OMWEA [OEMM UEWA]

According to the semantic analysis we have:

a. (Vehicle) [(celestial body) (change of three-dimensional framework)]

The vehicle [moves between celestial bodies by changing three-dimensional framework].

b. (Change of three-dimensional framework) [(vehicle) (celestial body)]. By changing three-dimensional framework of [the vehicle moves between celestial bodies].

c. (Change of three-dimensional framework) [(celestial body) (vehicle)]. The change of three-dimensional framework [allows movement between the celestial bodies of the vehicle].

CONCLUSION ON SENTENCES STRUCTURE

The sentences of the Ummo language at the first level are composed of words forming a conceptual proposal. The words of a sentence in this language are commutative, which means that they can be moved around but still give the same overall meaning.

CONCLUSIONS ABOUT GENERAL SEMANTICS

The foundations of the semantics of the language at the first level can be decoded and will help us to understand the documents. The meaning of the words is linked with the context and our understanding of the Ummo culture. The first level Ummo language is "phonetic-conceptual". It is a "Russellian" type language which has the following properties:

- The words are phonetically meaningful (and not "orthographically"), which we know because only one spelling is strictly meaningful for one phoneme, and not all spellings.

- The words are the result of the relationship between phonemes and concepts:

—The words are the result of the CONCATENATION OF A PAIR OF NESTED CONCEPTS (primary concept, established concepts, or propositional concept).

—An Ummo word is like a number. It must be read from LEFT to RIGHT, from the MAJOR concept to the nested MINOR concept. But for the analysis we must decode the word in the opposite direction: from RIGHT to LEFT.

—In EACH NESTED PAIR A MEANING EMERGES, which is related to the upper nested phoneme-concept.

—For TRANSLATING PURPOSES, the relationship between two phonemes-concepts can be decoded in English by using the word "get". This translation of the relationship through the use of "get" allows us to order the decoding so that we can perform the complete translation of the word.

*

How the exocivilization's word analysis allows revolutionary discoveries

One of the goals of my research of the Ummo language was to decipher the words so that I could obtain precise information on some important topics described in the letters. Here are two discoveries directly attributable to the decoding of the Ummo words.

A REVOLUTION FOR TERRESTRIAL AGRICULTURE

When decoding the word IXI, I came up with "amino acid" and got a big surprise. The decoded word seemed to have no meaning within the context of the terrestrial bio-chemistry. However the fact that the word IXIMOO means "protein" made it obvious that the word IXI was connected. . . Was my method of decoding inaccurate?

Of the 350 decoded words we had at the time, I decided to analyze the words that appeared close, along with their contexts:

D58-2	03/04/1967	ESP	IXI	The second C (ARN) carries a (IXI) amino acid (remember that amino acids are true modules or link of PROTEINS)
D57-1	23/01/1967	ESP	xii	FREQUENCY UNIT (In periodic phenomena): This is the "XII", whose translation has various meanings for this word is not only the DAY UMMO but also the rotation of a wheel or the cycle time per unit
D66	1967	ESP	IXINAA	various methods of recording and reproduction of IXINAA (sound)

I analyzed the concepts and my first shock was that it seemed to me that the words were associated with concepts that had no relationship to each other.

It's a given that the word "iksi" describes an amino acid and the term IXINAA defines a sound or an audio frequency. Even stranger, the word XII, which is really close to IXI, means something like "a unit of frequency". All of this seemed totally incoherent...

For IXI therefore the transcription would be as follows:

. Identifies a cyclical distance identified

. Identifies a frequency specific

THE GRAPH OF IKSI

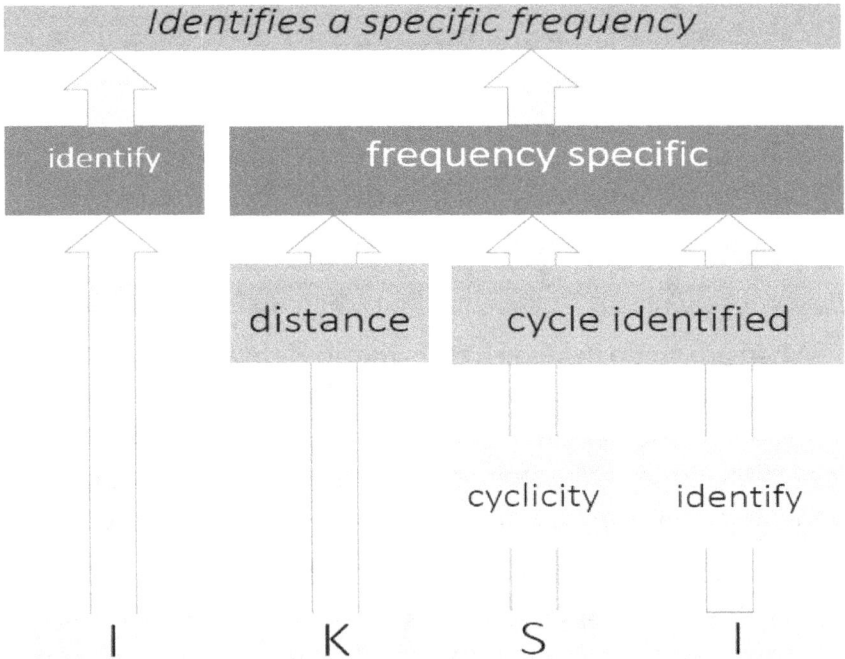

What relationship does this have with an amino acid? To try to figure out this we will next analyze the word IKSIMOO.

THE GRAPH OF *IKSIMOO*

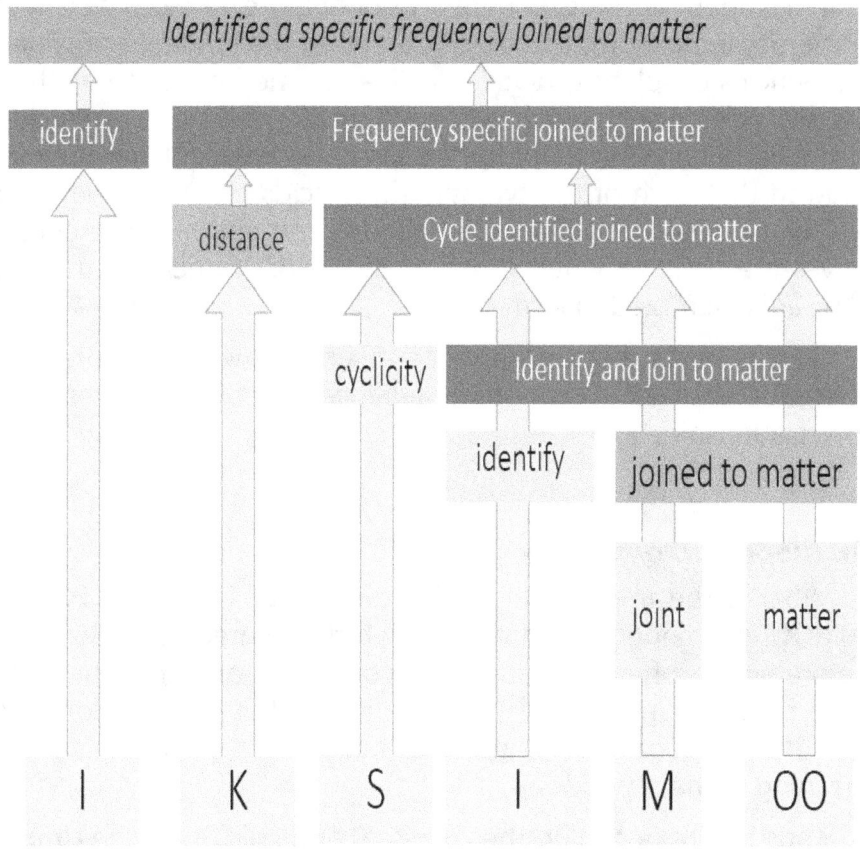

Given the contextual guidance, I could now do the translation as follows:

. Identifies the cyclical identified which joins matter

. Identifies the frequency specific and its materialization

. Amino acids and their materialization

In other words: amino acids constituting the protein. That part of the translation seemed perfectly clear.

But what connection could there be between an amino acid and the "identification of a frequency"? For six months I consulted with many biologists on and off the Ummo research team without result. Then I finally found the first clue...It was a thesis coming from a student at the Graduate School of Biotechnology at Ghent University in Belgium. This thesis described the influence of sound frequencies on plant growth. A colleague then gave me another decisive reference. It was the work of Joël Sternheimer, a researcher at the European University for Research in Paris. Joël Sternheimer showed that each of the twenty amino acids emitted a wave for which we could calculate the specific frequency. These sound waves are emitted when these amino acids, carried by transfer RNA, are assembled to form proteins.

So it appears that the separate meanings were not unrelated after all. Our society was just not advanced enough to understand this relationship.

AN HIDDEN INFORMATION

What is the implication of this discovery? The terms IXI and IXIMOO are mentioned in a letter which is dated 1967, 30 years before Mr. Sternheimer made his discovery. No-one on earth could have possibly understood the meaning of this word before the year 1997. As for me, I had no idea that I would predict this exact meaning of these words.

Thus by the decoding the words IXI and IXIMOO in the Ummo letters written 30 years ago, we have confirmed the revolutionary discovery that epigenetic modifications of plants may be possible to use the influence of sound frequencies.

The practical upshot of this is that in the 3rd millennium we may be able to magnify agricultural yields and performance without the use of dangerous chemicals or GMOs.

UMMO SPACESHIP LOCOMOTION TECHNIQUES

The technical vocable IDUUWII AYII is mysteriously presented without any precision concerning the "propulsion" equipment spread out inside a revolution toroid (*). This propulsion equipment seems to be located in a ring-shaped enclosure, built in the DUII (equatorial ring or crown surrounding the craft). The semantic analysis of this vocable, gives us indications about this type of "propulsion".

TRANSCRIPTION OF THE TERM "IDUUWII"

I	identification (oneness)
D	form, appearance, manifestation
UU	dynamic dependency (relative to the fields of forces)
W	generate, create, produce
II	limit, frontier, membrane

According to the coding of the Ummo vocables, the basic concepts complete each other from right to left. To attempt to properly define this term we need a chart which rigorously respects this organization:

IDUUWI is thus literally "the identification of the emergence of a dynamic dependency linked to the creation of a frontier". For this term to be more meaningful we could transcribe it thus: "identification of a force (attraction/repulsion) which generates a frontier effect" with which there is interaction. The term AYII is clearly identified as a "field".

The IDUUWII AYII technology is thus a system producing a field of dynamic forces enabling spaceship propulsion or levitation. What is disturbing in this "exotic" concept is the fact that it does not refer to any notion of the engine, combustion, rocket type propulsion or other. The idea of a generator of fields of forces creating an interaction with the environment or cosmos must be considered within the Ummo cosmological context which integrates a cosmos and an anti-cosmos with inverted properties. These two cosmoses are separated by a relay layer named XOODII. It's a membrane, a frontier which can be strongly distorted during wrinkling of space.

It thus seems logical to infer that their technology is sufficiently sophisticated and powerful to be able to modify, locally, the membrane which separates the cosmos from the anti-cosmos. We are speaking, indeed, of a technology which creates a force field on this membrane called XOODII, which, by reaction, under the pressure of the anti-cosmos, creates a force in return. This force in return is the anti-gravitational effect making levitation possible. This is indeed a technology which plays on the cosmos/anti-cosmos frontier for transportation or levitation on the same spot. It is a bit as if UFOs were playing trampoline with the inter-cosmic relay layer.

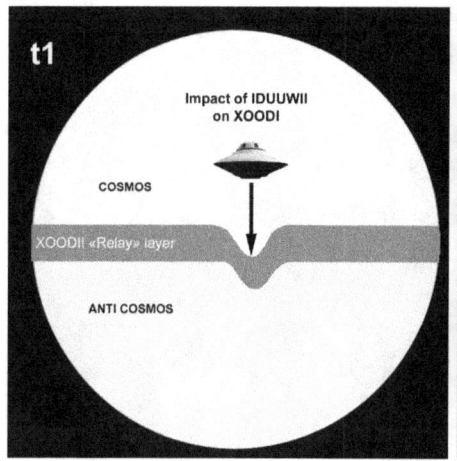

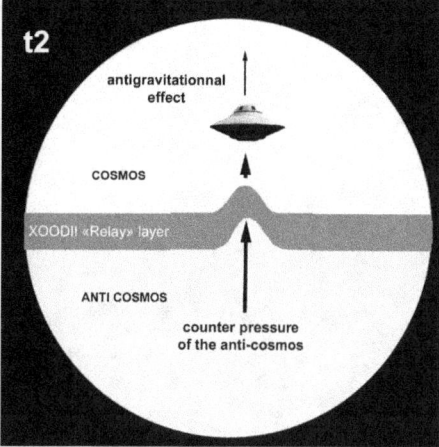

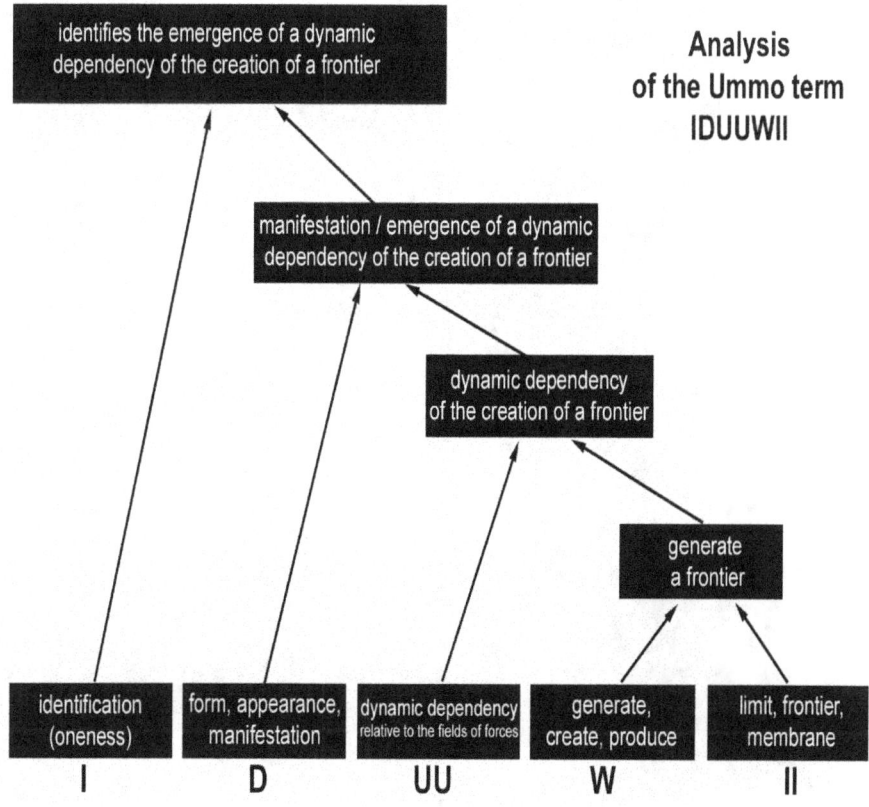

**Analysis
of the Ummo term
IDUUWII**

*

Learn to Count in Ummo language!

Presentation

On Earth man began counting by using his fingers. Because of this, most civilizations adopted a numbering system of base ten, and sometimes twenty if you included the toes. In very rare cases we developed a base sixty. Therefore the base of our numbering system was rooted in the physiological world of having 5, 10, or 20 digits. The few other systems which arose were due mostly to cosmological or theological symbolism. For ancient historical reasons which are unknown to us, the Oomomen use the base 12 system of numbering. The symbolism could be linked to astronomical-magical beliefs, which are some of the usual attributes of power in a primitive society.

D41-15: When the mind of the old hierarch OES 17 son of OES 14 dematerialized (dies), the young girl IE 456 is elected by the AASE OUIA Council (governors) met in the NAATOWSEE UA NAII valley (it was an ancestral custom to meet in the historical Valley to elect the Grand Dignitary of UMMO and twelve councilors).

D-541 ESP—10/04/87: We know that around 11.8 million Ummo-years (2.5 million Earth years), our forefathers wrote with almost ideographic alphabet and counted using the decimal system as you do now (Right now our number system is base 12). Even in year 1 of the first Era they had always decimal system. In year 1 it was believed that humanity would live for only six thousand Ummo-years (around one thousand two hundred and seventy years of the Earth years).

D 45—ESP – 1966: EXPRESSIONS WUA (MATHEMATICS) OF UMMO.

Our system COUNT is base 12; on Earth you chose a decimal system. Apart from the logic fact that signs used by Earthlings are different, expressions of complex numbers is done by ordering the numbers in the same way as you do. But one who is not familiar with this numbering system on base 12 that may seem strange to write a quantity of 29 trees for example, writing 25 in base 12. Any of your mathematicians will solve this apparent contradiction.

D 45—ESP – 1966: WUA (MATHEMATICS) EXPRESSIONS OF UMMO

Here we have added the table of basic signs and put them near the corresponding expressions.

You can assume that the enormous complexity of mathematical expressions, logic and geometric is solved (like you) by a large number of symbols that do not exactly resemble those used on Earth. We can note a curious fact: in your algebraic expressions you symbolize the numbers by letters. On UMMO we use a wide range of special symbols.

The Ummo numbering system has many features in common with other terrestrial numbering systems, while also being totally specific and unique. The symbolic representation of numbers is associated with a structured building of numbers from right to left, like our current numbers of Indian origin. The great similarity of these two numbering systems is both prodigiously astonishing and almost inevitable, considering that the means of developing complex systems pf operations leave few possibilities for the composition of the basic structure of a numbering system. Apart from the symbols of numbers 0, 1, 2, 3, 4, 8 and 12 which can be found in very different ancestral cultures on Earth, having, of course, different significations; the other figures are simple graphics which are as yet unreleased!

NUMBERS AND GRAPHICS

THE ZERO

The zero exists in Ummo count, as in all modern systems of counts. It is denoted> and is pronounced "oo"—"o" (UO) in English.

Réf. D	Date	Language	Word	Extract
D63	25/07/1967	ESP	UO	we call for it IAGAIAAOO UO because it was first detected
D74	??/03/1969	ESP	UO	XOODIUMMO UO
D74	18/03/1966	ESP	UO	the layer XOODIUMMO UO
D41-5	1966	ESP	UOUAMII	UOUAMII (MEAN NUMBER ZERO)

Knowing that in almost all terrestrial number systems words related to the numerical values have an "historical" and/or "graphic" origin, there is a high probability that we should consider that some of the figures and proper names may be meaningful and some may not.

The word "oo"—"o" means "first", that is to say, "NUMBER ZERO". Note that Oomomen have an ordinal count from zero then shifted by 1 compared to ours. In other words:

	Earth	Ummo	Earth	Ummo
Ordinal Numeration	—	first	first	second
Cardinal Numeration	0	0	1	1

The translation in the context, we have:

. [U O]
. [dependency "get" entity]
. Depending on the entity

As we have UWAAM meaning "anti-cosmos", therefore in this direction we can conclude for UO, "U entity": Anti-entity.

THE ONE

The "ONE" is pronounced "i"—"as" and is denoted: 1 = —

D63:"IASXOODINAA(IAS=1)formedbyahighlyelasticmaterialofthermalandelectricalconductivityverylow.Inside
arearrangedcapsules(YAAEDINNOO)containingadoseofthesamematerialcalledUYOOXIGEE(ceramicproduct)
forming the outer layer already mentioned [UOXOODINAA - 31]. «

D33-3: "This is the BUAWAA IAS (Soul n ° 1)"

"An atom of krypton UAXOO IAS (receiver n ° 1)"

D59: "It will take now you make a mental effort to achieve a psychological translation
so that whenever in physics we talk about SIZE, the picture of a SCALAR does not touch
your conscience instead of 'IOAWOO (ANGLE that the hypothetical radius vectors of
TWO IBOZOO UU form between them).

It is nonsense to ISOLATE, in an effort to mental abstraction, a IBOZOO UU to study it.
We COULD EXPRESS IN SPANISH, by translating the POSTULATE known to our physicists: IIAS IBOZOO UU AIOOYEDOO (THERE IS NO IBOZOO UU INSULATED) ".

D74: "We can distinguish nine XOODIUMMOO DUU OII (which can be translated as
'connected layers') that have very diverse geophysical characteristics. The disconti-
nuity between these layers is not sharp; there are transition layers of varying thickness."

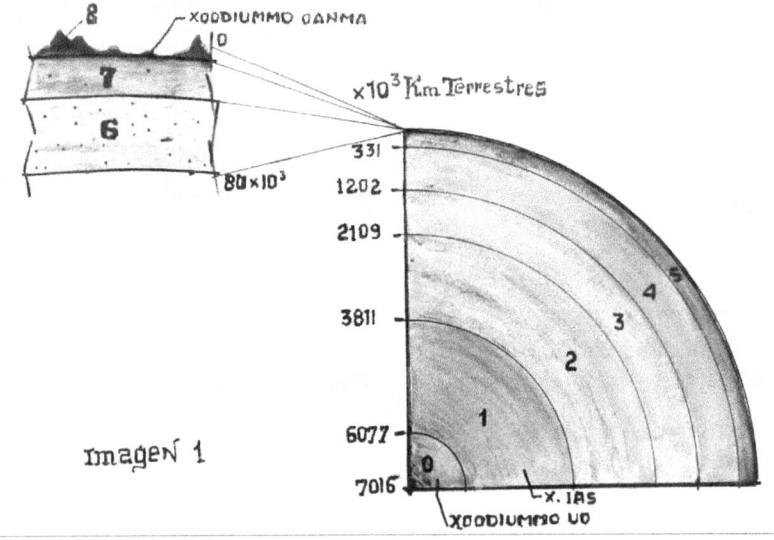

imagen 1

The picture 1 shows a section of our OYAA (planet) reflecting the
thicknesses of XOODIUMMO. The chemical composition of these
layers is varied. For example, the layer XOODIUMMO UO (NT: the
first layer of the center indicated "0")

"The top layer, the XOODIUMMO IAAS. . ." (Translator's note:
layer above the layer of UO, on the drawing noted X. IAS)

In this case the phoneme can describe the associated graphic (the
1 is often associated with a stick unit.) In other cases it will be a unique
property (e.g. Pi.) The stick graphic suggests that the formalism of the

figure ONE is very old. The phoneme "i"—"as" expresses the cardinal concept of the "unique value" that is to say the name of a figure 1, the number of value 1, and the number 1. But this phoneme does not express the ordinal, because Oomomen count starting at zero. The number 1 is therefore the second digit in the Oomo numbering system.

The translation in the context is:

.   [I [A S]

.   [Identification "get" [displacement "get" cyclicality]

. Identifies the displacement of a cycle

1 = Identifies an occurrence

THE TWO

The "TWO" is pronounced "i"—"en" and is denoted: 2 = Γ

D33-3	1966	ESP	IEN	The OEMII IEN (HUMAN BODY "2")
D33-3	1966	ESP	IEN	atom RECEIVER KRYPTON (UAXAOO IEN) IEN)
D59-2	06/05/1967	ESP	IEN	IBOZO UU IEN AIOOYAA (EXISTE). (IEN: pair, two)
D74	??/03/1969	ESP	IEN	These previous layers, solid, subjected to great pressure, are surrounded by the XOODIUMMO IEN ("2") and XOODIUMMO IEBOO (Nt: IEVOO identical to the D69-3 = "3")
D33-3	1966	ESP	IENXOODINAA	atom AAXOO IEN (transmitter "2")
D69-3	Juin 1968	ESP	IENXOODINAA	IENXOODINAA. (IEN = 2) This is a layer or submembrane crystalline silicon dioxide and modulated in the form of hexagonal mosaic.

Again the phoneme can describe the associated graphic (like the ONE often associated with a stick unit). The concept implies a cardinal number. It logically means the name of the figure 2, the amount of the value 2, the number 2 itself, and a pair. The number 2 is the third digit in the UMMO numbering system.

The literal translation is :
.   [I [E N]
.   [Identification "get" [concept "get" stream]

. Identifies the flow concept

The idea of a concept of "2" associated with a stream is a very original idea. Indeed, it is conceivable to define the concept of a process or stream, but to do so it is necessary to have at least two separate entities: X-stream—> Y.

So, when you express the concept of a "flow" it implies that you have 2 separate entities.

2 =Identifies the flow concept

THE THREE

The "THREE" is pronounced "i"—"ebo" and is denoted: 3 = Π

D74	??/03/1969	ESP	IEBOO	These previous layers, solid, subjected to great pressure, are surrounded by the XOODIUMMO IEN ("2") and XOODIUMMO IEBOO (Nt: IEVOO identical to the D69-3 = "3")
D69-3	Juin 1968	ESP	IEVOOXOODINAA	IEVOOXOODINAA. (IEVOO=3) (IEVOO = 3) It is the layer or bark of the innermost XODINAA.
D69-3	Juin 1968	ESP	IEVOOXOODINAA	These sensors integrated into the refrigerated IEVOXODINAA.

Again the phoneme is a graphic which is associated with ancient linguistic practices. The translation is:

. [I [E [B OO]
. [Identification "get" [concept "get" [interconnection "get" matter]
. Identifies the concept of an interconnection materialized

Knowing that the phonetic word "i"—"bo"-long denotes a "network node", that is to say a dimensional interconnection between "flows" inward and outward (See IBOO) one can consider that we can "identify a conceptual point" (BOO) in space as the interconnection of 3 axes. Hence the translation is:

3 = Conceptual identification of a point.

THE FOUR

The "FOUR" is pronounced "i"—"es" and is denoted: 4 = O

D59-2: "We then use a type of multivalent logic that our specialists call UUWUUA IES (TETRAVALENT MATHEMATICAL LOGIC) in which any proposal will adopt one of four values:"

D59-5 "in WUUA (mathematics)"

"However, we resisted the temptation to add a more refined mathematical reasoning with our WUUA WAAM because it would require, as we say in the report that you initiate prior to the bases of our UWUUA IEES (TETRAVALENT MATHEMATICAL LOGIC). In thus the scientific rigor of the concepts that we express is irreparably damaged."

Here the number refers to the phonetic conceptualized representation of a figure which is unrelated to ancient linguistic practices, and which more logically would have been a square. This necessarily conceptualized representation refers to something different than a simple numbering listing. The phoneme UUWUUA IES is translated by the Oomomen as TETRAVALENT MATHEMATICAL LOGIC. The phoneme UUWUUA means "LOGIC" and the phoneme IES expresses "TETRAVALENCE", which is to say having "four values".

The translation:
. [I [E S]
. [Identification "get" [concept "get" cyclicality]
. Identifies a cycle concept
. Circle

In this case the graphic of figure 4 is represented by a circle. So the number 4 has a conceptualized representation. Note also, that this concept of "circle" is completely different from the functional "rotation" XII.

. 4 = Circle

THE BASE TWELVE

The base "TWELVE" is pronounced "di"—"e"—"wooee" and is denoted: 12 = ≥

The number on this graphic represents:
. a dozen and zero unit, so 10 in base 12

. $(121 \times 1) + (120 \times 0) = 12$ in base 10

12= ⊇	21= ‒ ⓓ	29= ⌐ ⓑ	67= ⓓⓓ	91= ⓓ ⓓ
13= ⹀	22= ‒ ⓣ	30= ⌐ ⓣ	68= ⓑ ⓞ	93= ⓓ ⓑ
14= ‒⌐	23= ‒ ⓠ	31= ⌐ⓓ	75= ⓣⓝ	96= ⓞ ⊇
15= ‒⌐	24= ⌐ ⊇	32= ⌐ⓞ	77= ⓣ ⓑ	99= ⓞ⌐
16= ‒ⓞ	25= ⌐ ‒	33= ⌐ⓑ	80= ⓣ ⓞ	100= ⓞⓞ
17= ‒ ⓑ	26= ⌐ ⌐	34= ⌐ ⓣ	82= ⓣ ⓣ	101= ⓞⓑ
18= ‒ ⓣ	27= ⌐ ⓝ	35= ⌐ ⓠ	85= ⓓ ‒	105= ⓞⓑ
19= ‒ⓠ	28= ⌐ⓞ	36= ⓝ⊇	. . .	120= ⊇⊇
20= ‒ⓞ				

See also "Errors or IQ tests?"

The translation of DIEWEE:
. Form [Identification [concept [generates a model]
. Form of identification (the concept that generates a model [the numbering base])
. Form of (identification the concept that generates [the numbering base])
. Form of (graph [the numbering base])
. Formalizing graph numbering base

12 = Graph of the numbering base

We can note that the word for this number also expresses that the number is the numbering base.

SUMMARY TABLE OF NUMBERS NAME

Number	Graphics	Word	Number	Graphics	Word
0	>	UO	7	ⓐ	
1	—	IAS	8	ⓞ	OANMAA
2	⌐	IEN	9	ⓑ	
3	ⓝ	IEBOO	10	ⓣ	
4	O	IES	11	ⓓ	

Number	Graphics	Word	Number	Graphics	Word
5	D		12	≥	DIEWEE
6	⊃				

OPERATORS OF THE UMMO NUMBER SYSTEM

Now that you have learned to count in the language of the Ummo planet, we, of course, will suggest that you learn to calculate in base 12. The numbering system has all the necessary operators for the development of mathematics:

D45: Here are some examples of algorithms using real numbers (base 12).

Earth	Ummo
$31 + 46 + 3 = 80$ Sum symbol: (Equality symbol: ⌐	
$2 \times 4 \times 8 = 64$ Multiplication symbol: ∫	
$\dfrac{12}{3} = 4$ Division symbol: ∣	
$4^3 = 64$	

169

Earth	Ummo
Root symbol: 	

OTHER EXAMPLES OF FORMULAS	
Earth	Ummo
Constant e	
Constant π	
Sh U (hyperbolic sine U) = $\frac{1}{2}(e^{u} - e^{u})$	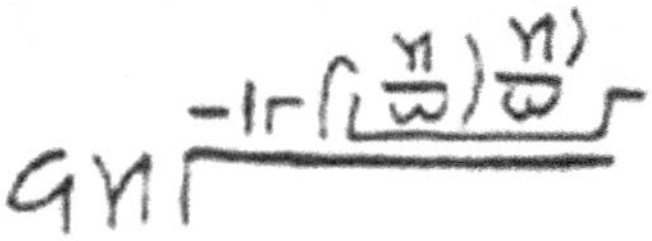
Expression of a delta (determinant) $\Delta = \begin{vmatrix} 3 & 2 & 0 \\ 1 & 5 & 2 \\ 0 & 1 & 7 \end{vmatrix}$	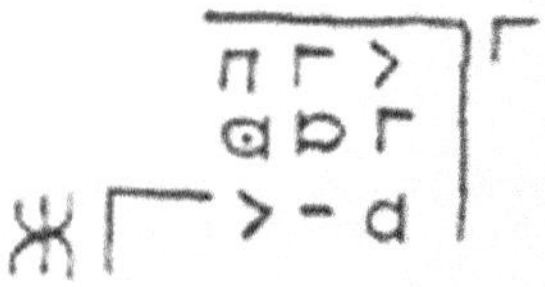
Two-dimensional matrix inversion	
Vector product $\vec{A} \wedge \vec{B} = \vec{C}$	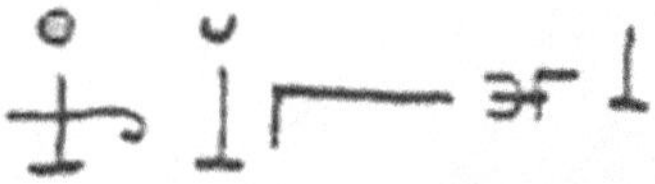
Derived $y = \frac{dx}{dy}$	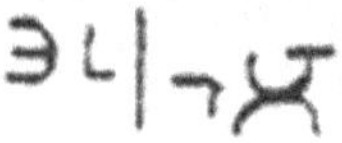

OTHER EXAMPLES OF FORMULAS	
Earth	Ummo
Tensor phi : Tensor ϕ	(ummo symbol)
Example of integration: $$\int_{0}^{0} Th \times dx = ln\ Ch \times \downarrow C$$	(ummo symbols)

FROM MATHEMATICS TO LOGIC

Gottlob Frege, Bertrand Russell and Alfred North Whitehead tried to show that logic creates mathematics. Indeed for the Oomomen, the debate between logic and mathematics is rendered a non-issue by the fact that mathematics (WUUA) depends on "Logic" (UUWUUA).

See "Decoding the words uua, wuua, uuwuua"

a. UUA = function
b. WUUA = generates functions = mathematics
c. UUWUUA = inseparable from mathematic = logic

We have the hierarchical nesting concepts as follows:

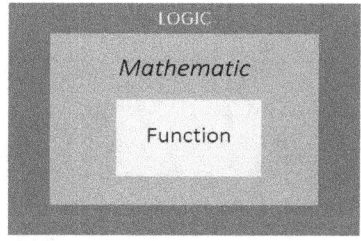

CONCLUSION ON THE UMMO NUMBERING SYSTEM

We have looked at just a few examples of the amazing historical and cultural coherence, and the extraordinary semantic consistency, of the figures which represent the Ummo numbers. The amazing consistency of all these elements is sufficient to betray an intelligence that goes far beyond the mere achievement of an anecdotal numbering system.

*

A REVOLUTIONARY LOGIC

The Oomomen use tetravalent logic, whose foundation was present on the Earth in the past. As our visitors from outer space say:

«...however, are not unknown to your thinkers and you will find early drafts in the Platonic literature and in the founding texts of Buddhist philosophy".

Tetravalent logic is also linked to the ontology, cosmology, and primary phonetic concepts of the language itself. We will explain these concepts and how they are linked.

Regarding ontology, "to be or not to be" truly is the question which we will answer.

To begin, let's pretend that only our physical world exists, and all other things including a 4th Dimensional realm, are unreal. These are just interpretations, a story we tell ourselves on a 4D world things definable in time, distance and volume.

The Absolute Reality, on the other hand, is probably "infinitely-dimensional" and inaccessible to us. You may recall, for instance, what Plato said in the "Myth of the Cave." A man had been trapped in a cave so that the only things he had ever seen were the shadows on the walls. For him, these shadows were reality. He would never see the Absolute Realty. Consider then that we are all like this trapped man in that we see a only a small piece of the Absolute Reality, and we interpret it based on that. Absolute Reality is what the Oomomen call AIODI.

With this in mind we will consider the interpretation of our eyes, our ears, and our brain of the physical world. This interpretation is what we commonly call "real." This interpretation may be what we call a stone, a car, a living being, a virus, etc. This is what the Oomomen call AIOYAA and associate it with the logical value of "TRUE".

Now, suppose I speak about a stone which will or could be in my hand. The stone of which I speak does "exist" (AIOYAA). But in fact there is no stone in my hand. This is what the Oomomen call AIOYAA EDDO and they associate it with the logical value of "FALSE".

TRUE and FALSE are the values that we use regularly, but they are strictly related to the physical 4D world. For thousands of years The Oomomen and also our own Indian philosophers have also presented two other values. This Tetravalent Indian logic is called tetralemne of Maya, or cosmic illusion.

In Tetravalent Logic the 3rd value is a conditional value. In some cases the result is "TRUE" and in other cases the result is "FALSE". For example, in quantum types of phenomenon the position of an electron is purely statistical. Sometimes the electron is there, and sometimes it is not. This is what Oomomen call AIOYAU and they associate it with a conditional existence.

The last value in Tetravalent Logic is very important to us in our everyday lives. Our own personal feelings, emotions, and interpretations of the physical 4D world exist in our heads outside of the physical world.

This existence is "TRUE", but only for US outside the physical world.

That is the "TRUE" of our interpretations of the physical 4D world, our feelings, our emotions because no one but I know what I feel or what I interpret.

According to Ummo documents, all of our neurological informa- tion housed in the brain is simultaneously stored in a cosmological receptacle which they call the "soul". In addition, every soul—which Oomomen call BUAWA—can store and also produce emotions and thoughts, which are processed in our brain and adapted to our environment. This is what the Oomomen call AIOYAA AMMIE and they associate it with the quality of being "unverifiable outside of the individual or collective conscious."

This "tetralogic" is the key to producing quantum conceptual leaps in many areas, including philosophy, and it may necessitate a complete rethinking and rewriting of western ontology. In the culture of the UMMO people, tetravalent logic, ontology, cosmo- logy, the primary phonetic concepts, and even the language itself form a coherent, homogeneous, and indivisible set.

THE TETRAVALENCE, LANGUAGE, PRIMARY CONCEPTS AND COSMOLOGY

In the Ummo language we have words that describe the concepts of tetravalence whose four values are:

. AÏOOYAA (existence verified)

. AÏOOYEEDOO (no reality outside any verification)

. AÏOOYAOU (real phenomenological potential or partially known)

. AÏOOYA AMMIÈ (unverifiable out of a field of consciousness individually and collectively)

D59: "As you know, formal logic accepts the criterion that you name "Principle of the third excluded" (every proposition is necessarily TRUE or FALSE). In our WUUA WAAM this assumption must be rejected. We then use one type of multivalent logic that our specialists call UUWUUA IES (MATHEMATICAL TETRAVALENT LOGIC) that any proposal will adopt four values:

AIOOYAA = (TRUTH)

AIOOYEEDOO = (FALSE)

AIOOYAU = (untranslatable in Earth language).

AIOOYA AMMIE = (can result: TRUE OUT OF WAAM)

NR20: "APPENDIX: GENERAL CONSIDERATIONS ON OUR FORMAL REASONING TETRAVALENT

We base our tetravalent system on the not formally accepted theory of the rejection of a mid term and a third term in the dialectic. In this system that which is NOT differing from the opposite of what IS. We accept that something can both BE AND NOT BE or NOT NOT BE NOR BE. It is certain that such ontological distinctions are rarely considered in everyday reality. However, they are not unknown to your thinkers and you will find early drafts in the Platonic literature and in the founding texts of Buddhist philosophy.

A and B are two sets of ontological realities enforceable in the dialectic, this system leads to acceptance of the four following combinations: X1 = {X □ A □ ¬B}, X2 = {X □ ¬A □ B}, X3 = {X □ A □ B}, X4 = {X □ ¬A □ ¬B}. You must translate here ¬Aet ¬B by "IS NOT A" and "IS NOT B".

None of the four forms of reality is the simple opposite of another.

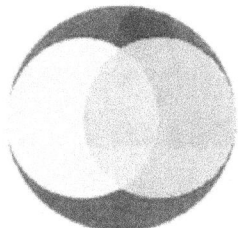

We give a simple example in the diagram below by considering the set of primary colors red, yellow and blue. Yellow represents the state (A) AÏOOYAA (verifiable existence) and red represents the state (B) AÏOOYEEDOO (unreal beyond a framework of verification). The orange color is a compound of red and yellow, the color blue is made neither with yellow neither with red. In this sense the complement of AÏOOYAA is not AÏOOYEEDOO. Thus the proposal contradictory for "X IS" and "X DOES NOT EXIST" is as follows:

—Reduced to a formal impasse Ø: (A □ ¬A) or (¬B □ B);

—Reduced to a phenomenological potential reality or partially known (A □ B). This state AÏOOYAOU is well summarized in the paradox created by your earth thinker Schrödinger Schrödinger which led to the deduction that two potentially contradictory statements could be superimposed due to the quantum nature of the phenomena implemented in the experiment.

—Extended to an existence AÏOOYA AMMIÈ (¬B □ ¬A), unverifiable out of the field of individual or collective consciousness, are at this level intellectual processes associated with abstract concepts or emotions such as empathy or compassion—which many of your thinkers associate with purely biochemical phenomena, and that we outsource partly to transcendent entities including the individual soul (BOUAWA), the collective psyche (BOUAWEE BIAEII), and God (WOA)."

THE FIRST VALUE OF TETRAVALENCE

The first value defined by the tetravalent word AÏOOYAA (verifiable existence) is linked to ontological positions of the primary concepts "O" and "OO", and also to cosmo-physical concepts that we will now explain in detail.

THE WORD AIOOYAA

The primary concept "A" is relative to the infinitesimal displacement of "angles-IOAWOO" of each IBOZOO because it is the basis of the Ummo physics. The primary concepts "AI" can be translated by the concept of "action".

There is a positive truth value for AIOOYAA when the network of a multrdimensional IBOZOO shows 4 of these angular dimensions. Anything sized angularly in a space-time is characterized by the infinitesimal displacement of "angles-IOAWOO" of each IBOZOO chained in each of the angular dimensions. If there is no "IOAWOO" in a space-sized angle, there is no verifiable existence. The limit of what is verifiable in sized space is ultimate "angle" IOAWOO which identifies the link between two IBOZOO of a string, following an "axis" OAWOO:

  - The movement identifies a materiality that has a dynamic spatiality

  - The angular displacement identifies a spatial materiality

The infinitesimal angle "IOAWOO" identifies the materiality of things (4-dimensional angular spatial, temporal) from 10D. In other words, this is: "the action to identify things materialized in the space-time" of 4D from 10D.

THE FIELD OF COSMOLOGY/PHYSICS

The documents express the concept of "truth" in our 4D cosmos (including time and space) "what is dimensional with characteristic of time and space."

D105: "the network of I.U. that is the AIOYAA [. . .] from various perspectives."

Briefly, AIOOYAA is what may occur in the spacetime proposed by Einstein/Minkowski. In other words, AIOOYAA is the network of

IBOZOO 4D-angles (a UXGIGIIAM WAAM), which occur from angular-10D.

D59-1, "a UXGIIGIIAM (SPACE) multrdimensional in its structure that undergoes multiple curves (we call weights), nothing like the concept EUCLIDEAN DIMENSIONAL SPACE."

D41: "dimensional (with characteristics of time and space)"

Anything sized angularly in a space-time is characterized by the infinitesimal displacement of the "angles-IOAWOO" of each IBOZOO chained in each angular dimension. If there is no "IOAWOO" in a space-sized angle, there is no verifiable existence. The limit of what is verifiable in a space is sized by the ultimate "angle" IOAWOO which identifies the link between two IBOZOO of a string, following an "axis" OAWOO.

THE PRIMARY CONCEPTS "O—ENTITY" AND "OO—MATTER»

(See also Chapter The primary phonetic concepts). The primary concepts "O—entity" and "OO—matter" are related to the cosmo-physics (which is my term for when cosmology and physics come together.) The Oomomen speak of "two classes of beings existing in the COSMOS" that are:

. The Neguentropic Living Beings (called AAIODII EXUEE)

They are seen in their 10 angular sizes and associated with the primary concept O = 10D multi-dimensional entity (including characteristics of time and space). Note that O is a multrdimensional entity in 10D which also has the 4D "characteristics of time and space". Confusion may occur because our Ummo friends speak of "dimensional" in the sense of 10D or 4D. It goes without saying that we tend to understand "dimensional" within the meaning of 3D or 4D.

. The Entropic Inert Beings (called AAIODII IOWAA)

They are associated with the concept OO = O "get" O. That is to say an entity that is 4D dimensional (having four angular dimensional of space and time) which translates to "matter" or "materiality".

CONCLUSION ON THE FIRST VALUE OF THE TETRAVALENCE

There is a positive truth value AIOOYAA when an IBOZOO multrdimensional network manifests 4 of the angular dimensions which translates as "The action to identify the things materialized in space-time".

THE SECOND VALUE OF TETRAVALENCE

The second value of tetravalence is defined by the word AÏOOYEEDOO, which means "no reality outside any verification".

THE WORD AÏOOYEEDOO

The phoneme "édo" expresses the concept of "MISSING—NOTHING—FALSE." For example "YAEYUEYEDOO (amnesia of fixation)"

Since there is no negation in the Ummo language, here we have the concept "no memory". In another example, "ASNEIIBIAEDOO (absorption by the BB) or disappearances" we also clearly see the concept of "NOTHING—FALSE." We also see this in a third example, EDDOIBOOI (WITHOUT WORK). Lastly, we see Nothing-False in the following quote, "On NOTHING we assign a verb that has no meaning for you; AIOYAYEDOO". The word AIOYAYEDOO then is concatenation (series of linked ideas) which springs from the concept AIOOYAA, which is dimensional and means "NOTHING—FALSE".

The overall expression EEDOO translates to "something" missing:
  . The model (or conceptual) has a material form
  . Conceptualizing a material form

The translation of AIOOYEEDOO = Moving identifies a materiality that has a spatiality that conceptualizes the material form.

We have: "Action to materialize things in a space that conceptualizes what is missing (and is therefore unverifiable)".

In other words, the "Action to conceptualize the absence of things in space-time".

THE COSMO-PHYSICAL AND PRIMARY CONCEPTS

Il y a une valeur de vérité positive pour AIOOYAA lorsqu'un réseau d'IThere is a positive truth value AIOOYAA when a network of multrdimensional IBOZOO expresses 4 dimensions of its apparent angular size. Conversely, AÏOOYEEDOO is the lack of expression of these four angular dimensions.

In this case, the concept of "OO—matter" is only conceptualized. Note that this conceptualization of OO is absolutely necessary to account for the absence of the "thing" matter called OO. In other words, one must have an idea of what there should have been to know that there is a lack of it. (See also Chapter The primary phonetic concepts).

CONCLUSION ON THE SECOND VALUE OF THE TETRAVALENCE

There is a truth value which is negative for AÏOOYEEDOO the "Action to conceptualize the absence of things in space-time".

THE THIRD VALUE OF THE TETRAVALENCE

The third value of tetravalence defined by the word AÏOOYAOU (real phenomenological potential or partially known).

THE WORD AIOOYAU

We understand from the word AIOOYAA, that the "action to realize the things in space-time" is "dependent", i.e. conditional or indefinite. The word "a"—"i"—"oyaoo" expresses the idea of "truth that is conditional or determinable."

"Reduced to a phenomenological potential reality or partially known (A ⊠ B)"

"This state AÏOOYAOU is well summarized in the paradox created by your Schrödinger which led to the deduction that two potentially contradictory statements could be superimposed due to the quantum nature of the phenomena implemented in the experiment".

THE FIELD OF COSMOLOGY/PHYSICS

Entities that have a quantum nature have a "reality" which is conditional or indeterminate based on the observation of the quantum event itself. This cosmological/physical event which happens in matter is due to an angular displacement of the "axis" OAWOO which follows the quantum event. These entities have a materiality which depends upon a statistical value, so their materiality is indeterminable.

CONCLUSION ON THE THIRD VALUE OF THE TETRAVALENCE

This is where the "Action to identify the things materialized in space-time is conditional (or indefinite)."

THE FOURTH VALUE OF TETRAVALENCE

The fourth value is defined by the word AÏOOYA AMMIE (unverifiable out of a field of individual or collective consciousness).

THE WORD AMMIE

In the Ummo culture an abstract concept such as a feeling or soul does exist outside of our cosmos, but it does "NOT EXIST" in the point of view of our 4D cosmos. Otherwise expressed the concept is: the non-existence in our 4D world, but an existence in another entity cosmological.

> "First we distinguish between two classes of beings existing in the UAANM (COSMOS) in contrast to two other major genres" OF THINGS (SERES) NOT EXIST.
>
> These are: AIOYAA AMMEIEE UAA [Such as WOAA (Generator), BUAUAA (Human Spirit), BUAWEE BEIAEII (Collective Spirit) or BUAUAA BAAIOO (Spirit Of The Living Being)]
>
> and AIOYAA AMEIEE OUEE (Such as the content of information, the sensation of pleasure, or a folk tradition).
>
> "True outside the WAAM [our universe]"
>
> "AÏOOYA AMMIÈ (¬B ⊠ ¬A), unverifiable out of a field of individual or collective consciousness."

In the word AÏOOYA AMMIE, the primary concept applies to AMMIE AIOOYAA. In other words, AMMIE applies the concept of the real 4D cosmos. In the word AMMIE, the primary concept of movement (A) applies in an inseparable way to identify a concept (IE), which is AIOOYAA itself in this case. AMMIE is translated to be:

. Moving inseparable from the identification of concept

. Move [out of our cosmos] inseparable from the identification of the concept [the concept AIOOYAA itself in this case]

THE COSMO-PHYSICAL AND PRIMARY CONCEPTS

(See also Chapter The primary phonetic concepts). The primary concepts "E—Concept" and "EE—model" are related to the fields of cosmology/physics. The Oomomen speak of two other major genres "OF THINGS (SERES) NO EXISTING", for which the primary tetravalent values and concepts are:

. The primary concept "E—concept" may be associated with any "concept" and in particular to the following concepts: AIOYAA AMMEIEE UAA designating cosmological entities that exist outside of our cosmos. Such as: WOAA (Generator), BUAUAA (Human Spirit), BUAWEE BEIAEII (Collective Spirit) or BUAUAA BAAIOO (Spirit Of The Living Being). These cosmological entities are not assignable to the primary concept "O" of 10D multrdimensional entities of our cosmos that include features of time and space. Those entities have no time and for us they can best be seen as concepts.

. The primary concept "EE" (meaning model) may be associated with the tetravalent value AIOYAA AMEIEE OUEE (Such as the content of information, the sensation of pleasure, or a folk tradition). These entities are linked to patterns in the cosmos called BUAWEE BEIAEII (meaning Collective Spirit), and designated by the term "BB".

CONCLUSION ON THE FOURTH VALUE OF THE TETRAVALENCE

That is: "Action to identify the things outside of our 4D space" in other words "something real in another cosmos".

*

11

1234567891011 12 13

GENERAL CONCLUSION

I worked very steadily for several years on about 1400 currently known pages of tex I worked very steadily for several years on about 1400 currently known pages of texts that contain 7,503 word entries (which are referenced online on my website) of a language totally unknown.

After over 40 years of mystery, I was able to finish deciphering the language of Ummo, which proved to be quite unique. I realized that the structure of the "words" was one of phonetic nested concepts, guided by functional logic. The meaning of each word emerges by the successive nesting of these primary concepts. This explanation is based on the results of the semantic analysis of literally thousands of words and word occurrences, using cross-comparison from a variety of differently dated documents that represent approximately 90% of all known Ummo words.

The language, number system, philosophical logic, mathematics and cosmology, all have an extraordinary consistency to the specific culture of the Ummo exocililization.

For most of us, to believe that there are UFOs have been a difficult intellectual test to overcome. To admit the existence of other extraterrestrial civilizations has seemed to be a large gap in our knowledge, while to realize the presence of some of them on our soil has been an abyss.

To quote the famous words of Sir Fred Hoyle, the probability that the UMMO file could occur by chance is equivalent to the probability of 'a tornado assembling a plane by scanning a dump". These results could not have occurred by either intuition or chance, but only through systematic logic. I feel that by translating this file I have uncovered the best evidence of the discreet, but active and peaceful, presence of exocililizations on our very own soil. Due to this translation, it is now possible to formulate hypotheses and explain new phenomena that were previously impossible because they were incomprehensible. I was led to this research by 'chance and necessity.' However it is now easy for me to conclude that Man is not alone in the solidarity of the vast and collective Cosmos.

In the next book to follow, I will present arguments on universal themes that are completely redesigned in a new paradigm. From the information of the UMMO file, we will then develop highly innovative theories on the origin of cosmological entities, the emergence of Life, the concept of the Soul, the influence of the stars on the psyche, telepathic communication, the emergence and the evolution of man, and much more. Many great philosophical questions which have haunted mankind for over 3,000 years will find plausible answers in response to the new paradigm.

Letter NR17 08/09/2003

'The voluntary discredit launched by the Intelligence Services and relayed by the media necessarily will yield beyond a certain threshold of credibility, however, it is higher than the simple logic would suggest. We see the same phenomenon of collective hypnosis and mental blindness with the multiplication of incursions into your airspace by UFO vehicles whose extraterrestrial origin should logically be no doubt. You will also find that some nations or geographic places focus the major part of the observation of these phenomena. This is not a coincidence but the result of a desire to exert some pressure on state agencies cunning and manipulators whose purpose is to maintain the social network under control to survive.'

For OUMMOAELEWE, 112 NABGAA daughter DORIO 34, approved by AYIOA son of ADAA 167

My personal demand of you is that you share the knowledge you have gotten from this book with as many people as you possibly can.

Then act in a way that you think is right to shift our planet, so that everyone can truly enjoy the Rights of Being Human.

Denis R. DENOCLA

'Knowledge for whom? Knowledge for what?'

ANNEXES

APPENDIX: LANGUAGE

COMPARISON OF METHODOLOGIES

In order to show the work that was done by Mr. Pollion and also compare the results obtained through the Denocla method, I am providing you with two examples of the translations done by Mr. Pollion:

BUAWUAA: This word appears twice. Here is an example of the definition of this word in one the Ummo letters: "the BUAWUAA (Soul) is not capable of processing data, to think, to develop information, but only to maintain, store, (grafted in a WAAM without 'dimension')". As translated by Mr. Pollion, the segment BUA evokes "contributions (B) mandatory requirements (AU)", that is to say "necessary contributions." The segment WUAA expresses "variations, changes, events, novelty, information (W) mandatory required (AU), truth, action, effectiveness (A)", that is to say "the information actually needed." In Mr. Pollion's translation, therefore, the whole word means "the information actually required mandatory contributions".

BUAUAA: This word occurs eleven times. Examples of definition of this word in a letter: "In a first language interpretation, and the phenomenon BUAUAA ontological entity or factor that attempts to represent or codify could be translated into the EARTH language as: AME; SPIRIT; PSYCHE or VITAL MOTOR". As translated by Mr. Pollion: the segment repeat UA in UAUA invites reading "plural AU", which expresses "mandatory need". B expresses the concept "contribution, participation". At the end the concept expresses, "truth, action, effectiveness". In this translation of Mr. Pollion's the whole word means "effectively mandatory contributions and multiple"

Afterwards there were twenty other translations done by Mr. Pollion for multiple and various "spellings": BUAAWA, BUAAWAA, BUAUAA, BUAUAAA, BUAUUAA, BUAWA, BUAWAA, BUAWAAA, BUAWUA, Buawuaa, BUAWUUA...

Work of Mr. DENOCLA

Please note that my work in semantic analyses was carried out methodically on a database of 7,503 word entries. Through this method I found that the structure of the "words" was much more complex than Mr. Pollion thought it was. The words were revealed to have a hierarchical nested structure, based on strictly phonetic concepts. Unlike Mr. Pollion, I was able to develop a systematic method of decoding which proved polysemy, or diversity of meanings, that depended upon the exact "spelling" of multiple "words". Through this method I was able to get a single significant word.

Using this method, I actually found 25 different "spellings" in the Ummo documents: BUAAWA, BUAAWAA, BUAUAA, BUAUAAA, BUAUUAA, BUAWA, BUAWAA, BUAWAAA, BUAWUA, Buawuaa, BUAWUUA, BUAWWA, BUAWWAA, BUUAUA, BUUAUAA, BUUAUUA, BUUAUUA, BUUAUUAA, BUUAWA, BUUAWAA, BUUAWAAA, BUUAWUA, BUUAWUAA, BUUAWWA and BOUAWA in French. However, all of these provide only a single phoneme. This phoneme can be transcribed by the significant orthographic form nested: B[U[A[WA].

Some references in the documents:

D105: "In a first language interpretation the phoneme BUAUAA and ontological entity or factor that it attempts to represent or codify could be translated into the language of the EARTH by: SOUL, MIND, Psyche, or VITAL MOTOR.

NR20 (French): 'our BOUAWA (soul)'

Following the Denocla method of decoding: (B) Interconnection 'get' [(U) dependence 'get' [(A) displacement 'get' [(W) generation 'get' (A) displacement] therefore the BUAWA is as follows:

- The interconnection depends on the movement generator

- More simply: Interconnection generator

Through the use of my method, therefore, I arrived at a single unified definition for any specific phoneme, which was impossible to do through the use of any previous method.

(See detailed analysis of the Extracts from the 'Ummo language Dictionary' D.R. DENOCLA)

EXCERPTS FROM THE 'UMMO LANGUAGE DICTIONARY'

To perform a comprehensive analysis of the Ummo 'words', I needed to build a database of all known occurrences along with their references and their context. Half a dozen people also joined in this extensive work. Thanks to this collective work, by the year 2006 this database contained 7,503 entries with direct access to semantic analyses on 299 meaningful words. At that time there were about 5% of the words that were still awaiting analysis. Then, word for word, I combined their analyses to instances in the database and to the 'Ummo language Dictionary'. I will use a few particular words to help you understand my process.

SEMANTIC ANALYSIS OF PHONETIC WORD 'OOM'—'MO'

The context

D110	19/05/1969	ESP, prov. Australia	OOMO	They also inform me of your delusion, logic, referring to the announcement possible the arrival of a nave of Oomo
NR-18	19/12/2003	FR	OOMOAN	We use interchangeably in order of preference and the terms ooman, oomoman, oomoan in our correspondence with your English speaking brothers.

D 84	04/09/1969	FR	OUMMO	We designate our planet with a phoneme that you might transcribe as follows: OUMMO.
D 379	05/02/1988	ESP	UMMMO	We come from a planet whose name phonetically expressed in Spanish is UMMMO.
D60	27/05/1967	ESP	UMMO	from our OYAA (PLANET) UMMO
	30/01/1988	ESP	UUMMO	from a planet whose name expressed phonetically can be verbalize in Spanish: UUMMO.

D21 (Esp): UM-MO (the 'U' very close and guttural, the M could be interpreted as a B)

D70 (Esp) published in Paris—dictated by XOODOU-7

The words marked with (*) are in French in the original text and several elements prove the origin of a French typist (Rivera instead of Ribera, for example).

«... Oumo ("m" extended to the pronunciation) . . .'

Identifying concepts

Ummo is a "proper name" which is the planet of Oomomen "OYAA (PLANET) UMMO"

Syntax identification

So we have:

ESP— Prov. Australia OOMO

FRA OUMMO

ESP UMMMO

ESP UMMO

ESP UUMMO

The significant syntax used is UMMO.

Identification of long sounds

Th written word "UMMO" should read "oom"—"mo".

The significant syntax "MM" describes the "dynamics".

The transcript of the term UMMO

see of the table primary concept

Following the Denocla method of translation, we build functional and conceptual meaning:

[(MM) inseparable "get" (O) entity] = inseparable of the entity

(U) dependence "get" [(MM) inseparable "get" (O) entity]

= link inseparable of the entity

UMMO = 'link inseparable from the entity "

Therefore: OYAA UMMO = planet of "link inseparable from the entity"

This makes even more sense when we consider that the BB-planetary cosmos is inseparable from the planet and living beings that inhabit it. It is logical that we could then translate this term as "link inseparable from the living beings that inhabit the planet."

SEMANTIC ANALYSIS OF "OO"—"E"—"WA"

The phonetic phoneme "oo"—"e"—"wa" means the concept vehicle, vessel.

Identification of syntax

D 84	04/09/1969	FR	OUEWA	At 4:17 am mn GMT Earth's day March 28th 1950, one OAWOLEA OUEWA (lenticular spacecraft) established contact with the lithosphere
D74	??/03/1969	ESP	UEWA	At 4 hours 17 minutes 3 seconds GMT Earth's Day March 28th 1950, a OAWOLEA UEWA OEM (lenticular Spaceship)
D57-1	04/09/1969	ESP	UEUA	Twenty-four men went to this Planetary System within two OAUOLEEA UEUA OEMM (This is how we call vehicles with lenticular shape moving outside our atmosphere).
		ITAL	UEUAA	Where was injured a OEMII from Earth, victims of our UEUAA OEEMM (Spaceship) and left for dead.
D41-11	1966	ESP	UEWA	in any UEWA (vehicle)

191

D69-2	Juin 1968	ESP	UEWAA	Including phonetic roots: OAWOO = size; OOLEEA = penetrate, pierce, = UEWAA vehicle, vessel;
D 731	Juin 1968	ESP	UEWUA	UEWUA (spaceship) does not invert their IBOZSOO UHU in the direction (-M)
D37-2	Févr. 66	ESP	OMWEA	. . . took place in the OAWOOLEA oemm OMWEA UMMO 56...
D41-6	1966	ESP	NOIA UEWA	these ancient NOIA UEWA

We have therefore the following spellings:
FR OUEWA
ESP UEWA
ESP UEUA
ITAL UEUAA
ESP UEWA
ESP UEWAA
ESP UEWUA
ESP OMWEA

OMWEA spelling is a distortion due to the fact that the phoneme was heard and wrote "omwea" instead of "oo"—"e"—"wa" Given the letter D84 in French, I get the significant form: UEWA.

Identification of ambiguous sounds

We retain the ambiguous sound "wa" with the syntax "WA".

Identification of long sounds

It remains to identify if the final sound "a" is long or not.

We need to use the semantic analysis to try to identify the meaningful spelling.

Transcription of the word UEWA

Depending on the conceptual method, we build:

(U) dependence "get" [(E) concept "get" [(W) generation "get" (A) movement]

Relationships between phonemes	Transcript relationship between functional concepts	Synthesis
W "get" A	[(W) generation "get" (A) displacement]	Generation get displacement
E "get" WA	(E) concept "get" [(W) generation "get" (A) displacement]	The concept of the generation of a displacement
U "get" EWA	(U) dependence "get" [(E) concept "get" [(W) generation "get" (A) displacement]	Depends on the concept of generating a displacement

A transcript of UEWA is: Depends on the concept of generating a displacement

The graph of UEWA

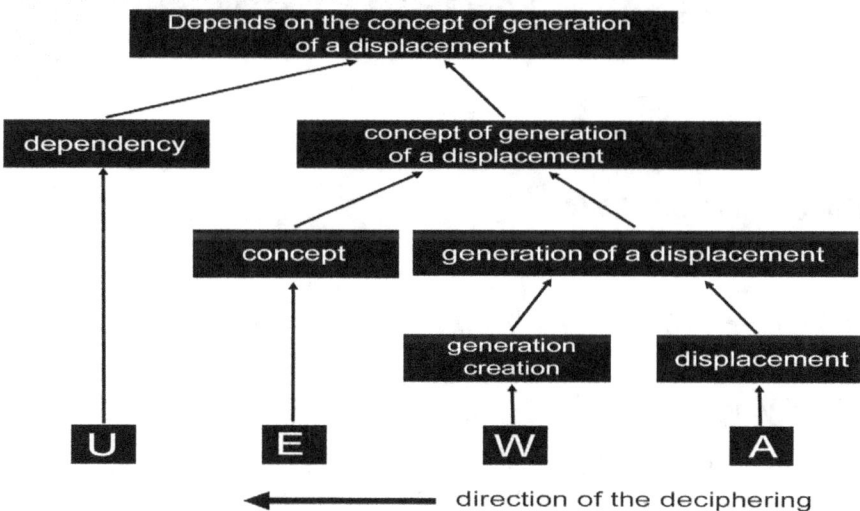

The translation of the word UEWA

In the context of translation, this word is simple to understand and we are very familiar with it:
  . Depends on the concept of generating a displacement
  . Depends on the (concept of the generation of movement)
  . Vehicle

UEWA NOIA

Conclusion on UEWA

The syntax UEWA describes the concept of "vehicle".

SEMANTIC ANALYSIS OF PHONETIC WORD "WOO-AM"

Identifying concepts

- . the WAAM means our "cosmos"
- . the UUWAAM means our "anti-cosmos"
- . the WAAM-WAAM the means "multi-cosmos" consisting of all pairs of cosmos.
- . the WAAM-U means the "cosmos of Individual Minds' "
- . the WAAM-UU (also called BB-global or WAAM [OU]) is the "cosmos of "Collective Minds" and modeler of the cosmos"

Spelling identification of WAAM

| D59-2 | 06/05/1967 | ESP | UAM | NO, ABSOLUTELY NOT: Our picture of the UAM (COSMOS) that is to say, the SPACE |
| D105-2 | 12/07/72 | ESP | UAMM | It is possible, using technical means, to pass from UAMM to another UAMM. |

D105-2	12/07/72	ESP	UUAMM	Us, we know there is a UUAMM (Not accessible by technical means) in which a hypothetical OEMII that could move within it (completely absurd hypothesis)
D105-2	12/07/72	ESP	UUAMMM	Inside the UUAMMM as we mention, you can consider that the SPEED OF LIGHT, measured within it, is INFINITE.
D 33-1	18/03/1966	ESP	UWAMM	the assumption of WAAMWAAM (PLURICOSMOS) it is because we observe that in our UNIVERSE and in UWAMM (COSMOS complementary opposite electric charge) There was a very small number of existence possibilities
D 792-1	janvier 1988	ESP	WAMM (OU)	The disturbance caused in BB results by a frontier effect, WAMM UU eliminates responsible for the disruption.
D357-1	12/03/1987	ESP	WAM-WAAM	This does not make sense to speak of "CENTER" in the WAM-WAAM.
NR-13	14/04/2003	FR	OUWAAM	because of poor isodynamic conditions space whose folds, generated by our OUWAAM, allow to make our trips with durations of course acceptable.
D62	4–5-6 1967	ESP	UAAM	the existence of other civilizations on planets of UAAM (UNIVERSE)
D105-1	12/07/72	ESP	UAAMM	Under this concept, the end of UAAMM (cosmos) would not be as assumed by some Earth Cosmologists, "a terrible sea of thermal radiation," but rather a bewildering world in which only the TIME dimension as, have his home
D63	25/07/1967	ESP	UUWAAM	curvatures of the three-dimensional space are caused by the UUWAAM (COSMOS TWIN)
D41-15	1966	ESP	UWAAM	our two twin cosmos, WAAM (ours) and UWAAM (his twin)
D105-2	12/07/72	ESP	UWAAMM	Yet (insurmountable contradiction bases with binary logic) there is a UWAAMM wherein the speed of light, in the absence of disturbance mass, will be infinite.
D 792-1	janvier 1988	ESP	WAAM	These can cause folders in the space-time universe environment, or transfer mass and energy and also the information through
D 731	20/03/1987	ESP	WAAM (U)	All macro-organism has a soul: The WAAM U which individual psyches ultimately modulate the structure of WAAM (OU)
D357-2	12/03/1987	ESP	WAAM (OU)	WAAM (OU) IS THE MODEL OF THE WAAM-WAAM

The spellings are numerous, to adjust for the overlap we will use the significant spelling "WAAM".

DECODING the word WAAM

So we have (See Rule of Semantics):

(W) generation "get" [(AA) dynamic displacement "get" (M) Joining]

So we have a "join" of (W) and (AA), that is to say a generation dynamically and simultaneously.

WAAM = generation and dynamic simultaneous

CONCLUSIONS

The WAAM-UU has a dynamic dependence on all the cosmos as a pilot using his models.

The WAAM-U has a link with all of the BUAWA.

The UUWAAM is the cosmos of which our WAAM (as a reference point) depends on.

*

APPENDIX: Norwegian Cargo

Preamble

In «Presence 1, UFOs, Crop Circles and Exocivilizations» we saw in the analysis of the Chilbolton Crop Circle, that our visitors had detected the radio signal from the Arecibo antenna, thanks to a fractal geometry interferometer-type device.

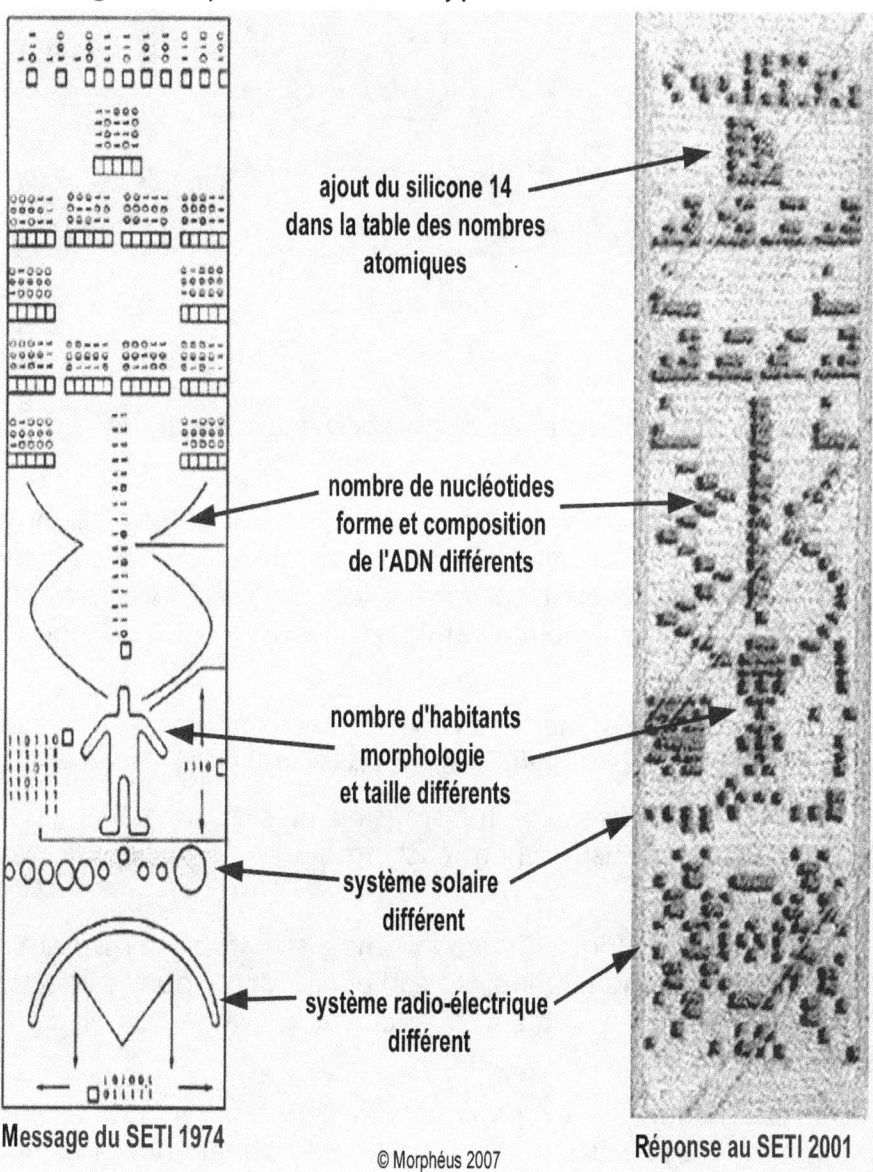

ajout du silicone 14
dans la table des nombres
atomiques

nombre de nucléotides
forme et composition
de l'ADN différents

nombre d'habitants
morphologie
et taille différents

système solaire
différent

système radio-électrique
différent

Message du SETI 1974

© Morphéus 2007

Réponse au SETI 2001

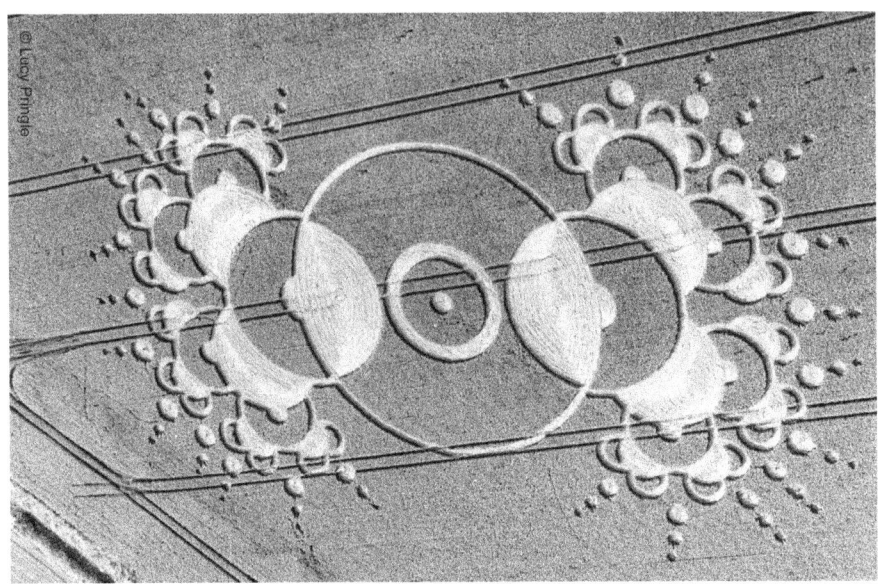

1. The role of interferometry in compensating for signal weakness

The main problem with the signal of the Norwegian freighter in 1934 is that after traveling 14 light-years, it would be incredibly scattered and drowned in cosmic background noise. A conventional radio telescope, even a gigantic one, would not be enough to capture it.

In the early 1930s, radio transmission technologies were still in their infancy for very high frequencies (VHF and UHF).

• The tubes used for transmission at that time had poor performance, especially when reaching high frequencies such as 413.44 MHz.

• This technological limitation means that the transmitters lacked precision and filtering. Instead of emitting a pure, perfectly targeted signal on a single frequency, the emission «drooled» or spread out on adjacent frequencies.

• In addition, these rudimentary transmitters generated many harmonics (frequencies multiple of the basic frequency). It is not impossible that the original emission of the cargo ship took place at lower frequencies (around 137.5 MHz or 206.5 MHz) and that it was in fact the 2nd or 3rd harmonic that was captured in space.

• In the context of the interception of a very weak interstellar signal, the fact that the original signal is «imperfect» (spread out and rich in harmonics) paradoxically becomes an advantage for a civilization with adequate listening technology.

Interferometry consists of using an array of multiple antennas spaced from each other. By combining their signals, a virtual telescope is created whose size is equivalent to the distance between the most distant antennas (this is how the Earth was able to photograph a black hole).

If the UMMO/IUMMA civilization had a planetary interferometric network, or even extended to its solar system, the global collection area and resolution would become sufficient to isolate an extremely faint artificial signal from cosmic noise.

It is very likely that technical improvements to this type of device will be possible, and allow this detection performance more easily...

2. The contribution of fractal geometry

Integrating fractal geometries into these interferometric antennas would add two major technological advantages that fit perfectly with the detection of an unexpected terrestrial signal:

• Ultra-Wideband: Unlike conventional antennas that are generally optimized to pick up a single specific frequency, fractal antennas have the ability to resonate over an extremely wide frequency spectrum. This helps to explain how such a system was able to pick up «by chance» the very specific frequency of 413.44 MHz emitted by the cargo ship, without the need to point at a receiver previously set to this specific band.

• Compactness and efficiency: The fractal geometry makes it possible to maximize the reception perimeter while maintaining a very small volume, which greatly optimizes the capture of scattered electromagnetic waves.

3. Consistency with the Oommomen narrative

In the first document, Javier Fraile concludes that «it is nevertheless not forbidden to try to imagine future technological advances that would allow such a feat». The fractal interferometer corresponds exactly to this type of advance:

- It does not violate the laws of physics (the speed of light and the Doppler effect remain respected).

- It circumvents the hardware limit of the «maximum sensitivity» of a single antenna by relying on the massive processing of data (phase correlation) from a myriad of ultra-sensitive receivers.

In conclusion

The hypothesis is very solid in terms of physical theory. The technical imperfection of the Norwegian freighter's transmitter (which dispersed its energy over a wide spectral spread and multiple harmonics) increased its chances of being detected. A conventional radio telescope looking for a single frequency could have missed this signal drowned in cosmic noise, but a fractal antenna interferometer, thanks to its ultra-wideband capability, was able to capture all of these «leaks» (base frequency, spread and harmonics), thus accumulating enough energy and data to isolate and identify the artificial signal.

In order for an alien civilization to pick up a terrestrial radio leak of a few kilowatts dating back to the early days of our UHF technologies, the use of an ultra-wideband passive listening system, such as a giant interferometer equipped with fractal antennas, does not violate known laws of physics (such as the speed of light or the Doppler effect) and makes it possible to circumvent the hardware limits of sensitivity of a single antenna through massive data processing. To capture a terrestrial radio leak of only a few kilowatts dating back to the early days of UHF technology, the use of an ultra-wideband passive listening system (a giant interferometer equipped with fractal antennas) represents one of the few viable technological explanations.

Bibliographical references

http://ummo-sciences.org/activ/art/art20.htm

http://ummo-sciences.org/activ/art/art10.htm

APPENDIX: THE MYSTERY OF BINARY TABLES

Preamble

As I was bored, I had decided to spend 15 minutes on a riddle submitted by our U. visitors in December 2014, and which in fact was never solved...

In the end, I spent more than 15 hours and I made a small calculation error, which my friend Gemini is the only one to have seen. So here's the exact final solution.

PROBLEM STATEMENT

W1-113

@quark67 @MJUMBE4 @marsu35 @jpazelle @ostralopithec @Denocla Do not transcode. Transpose. Non-binary answer.

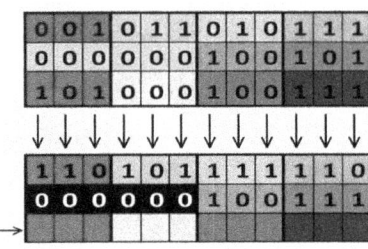

@MJUMBE4
Fortunately not. We only seek advice from our more advanced brothers on specific matters. In the absence of opposition we act freely.

@marsu35
Around 4 to 6 months. We intensively learn our assignment language during the journey to OYAGAA.

@jpazelle @ostralopithec
Your main military powers know of our presence and intentions perfectly. An official rapprochement was attempted several decades ago. The exclusively military aims of your leaders quickly rendered it obsolete. Unofficial relations remain within multi—national military organizations, such as NATO, with unscrupulous visitors. A livid, ophidian-like skin, and a hissing language earns them the erroneolous label of reptilian beings.

@DENOCLA
We cannot effectively work against this undesirable interference within your social network. Any repressive act towards them would have serious consequences. These OEMMII have a legitimate claim to interact with **your** your military authorities. They cause you no direct harm in the strict sense of the term... They are not the hostile people mentioned.

To start at the beginning...

We can see that the enigma is presented without context.

It could also have been broadcast in a separate message, as it is clearly unrelated to the responses addressed to the 6 people. It seems that there is still a desire to submit the enigma to these 6 people precisely.

Why?

The 2 tables have 12 squares. Why?

There is an apparent contradiction, the tables seem to contain a binary number, while in the instructions we are told «Non-binary answer. «.

Why?

Let's clarify the guidelines

Transcode: Translate into a different code. Synonyms of transcoding translating, transcribing

Transpose:

Place by inverting the order. ⊠ Swap.

To change form or content by moving into another field.

Summarize the guideline : Do not transcode. Transpose. Non-binary answer. ; tells us not to translate the boxes in Table 1 into a different code for Table 2.

But to place the content of Table 1 in another order in Table 2.

Background analysis of Table 1

We can place Table 1 in a particular context.

Letter D21 (1966)

INSTRUCTIONS

If someone a stranger tries to establish a verbal relationship with you (telephone or interview), it is not necessary to adopt a reluctant attitude to avoid any type of prank or to avoid the abnormal conduct of mentalities that present a supposed psychopathological syndrome.

Generally, pranksters present their arguments in a form that is not very elaborate. On the other hand, there are many forms of delusions whose structure, apparently perfectly coherent and logical, has a very high persuasive power, especially if the individuals who present them are intelligent patients (in the case of terrestrial paranoia, this syndrome is perfectly compatible with a high intellectual coefficient).

If in this case the person who claims to be a space traveler offers a minimum of coherent and apparently serious arguments, give him or read him this series of numbers (taking the risk that his identity will be usurped).

If he understands correctly and without you asking him, (you can simply say that these numbers were given to you by a friend, or better yet, that to prove his good faith you want to make him take a test) he will in turn have to give you a different sequence of numbers (also in binary system) that we ask you to NOTE.

In due course, we will give you the name of a person who will act as a liaison between you and us. If, we repeat, a case occurs (unlikely as far as you are concerned but statistically possible for any member of your group) I ask you, on behalf of this advanced group of UMMO, to write to us with the following information:

THE FIGURES DICTATED BY THE SUBJECT — THE IMPRESSION HE MADE ON YOU — THE VARIED DATA HE BROUGHT YOU — THE DATE OF THE CONTACT

Finally, we give you the series of numbers twice, repeated several times for greater security, lest multicopying make one of them illegible:

1 0 1 1 0 1 0 1 1 1 - 1 0 0 1 0 1 - 1 0 1 0 0 0 1 0 0 1 1 1

1 0 1 1 0 1 0 1 1 1 - 1 0 0 1 0 1 - 1 0 1 0 0 0 1 0 0 1 1 1

Letter D89 (1969)

Today, we transmit to you the coded text that we beg you to send to your strange communicator Francisco Atienza from a planet, unknown to us, URLN, as soon as a favorable situation arises for this oral, telepathic, ideographic communication or transmitted by any means deemed technically appropriate.

1 011 010 111 100 101 101 000 100 111 - 110 101 111 110 0100111

Several other documents confirm that this codification, which corresponds to Table 1. It is designated as binary, contrary to the instructions of the riddle.

Why?

Its use is clearly that of a key to identifying ET individuals. These are supposed to be able to understand the binary sequence and «it will in turn have to give you a different sequence of digits (also in binary system)».

This binary sequence is proposed to identify Saliano and Atienza.

Important: The binary sequence is not specific to an ET race, let alone linked to the star HD23065.

A priori, an answer by Saliano or Atienza could be Table 2, which is partially given to us, for our own evaluation.

In summary, Table 1 is a code understandable by an ET who will validate its ET origin by giving a similar answer to Table 2.

(a) If it is a transposition of the boxes, Table 2 would be a binary sequence, such that the boxes in Table 1 are placed in another order in Table 2.

(b) If it is a Transposition of the binary sequence, Table 2 would be a binary sequence, such that the figures in Table 1 are placed in another order in Table 2.

Analysis of hypothesis 1 Transposition of the boxes

Let's move the boxes with cut and paste

Table 1

too many

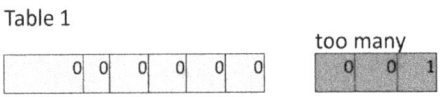

Tableu 2

too many

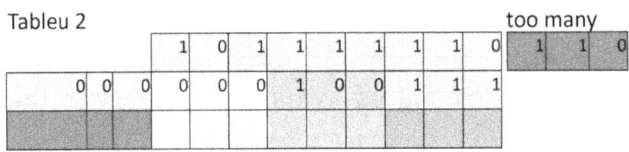

Max. 4 cells transposed

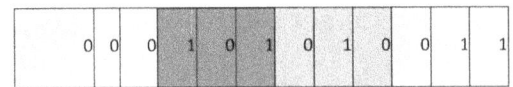

Conclusion:

•The transposition of the boxes is impossible, this hypothesis is not a possible solution.

•Transposition is necessarily a binary transposition of digits with 16 digits 1 and 14 digits 0.

Analysis of Hypothesis 2 Transposition of the Binary sequence

Table 1

0	0	1	0	1	1	0	1	0	1	1		16 digits 1
0	0	0	0	0	0	1	0	0	0	1		
1	0	1	0	0	0	1	0	0	1	1		14 digits 0

Table 2

1	1	0	1	0	1	1	1	1	1	1	0	13 digits 1 3 missing digits 1
0	0	0	0	0	0	1	0	0	1	1	1	5 digits 0 9 missing digits 0

For this assumption, the result line in Table 2 should have:

• 3 digits 1

• 9 digits 0

This gives us a very large number of possible answers.

This is clearly incompatible with an answer that would identify an ET...

Conclusion: the answer cannot be a strict transposition of the binary sequence in Table 1.

Analysis of hypothesis 3: the binary response

We have seen that there is confusing information about the possibility of a binary response.

This is indeed a binary encoding, but it is in fact necessary to understand that the binary sequence does not represent a single binary number.

It is the decomposition of the binary sequence into binary triplet squares that we need to solve.

The NUMICON Code Hypothesis

OYAGAA AYOO YISSAA 🔒
@Oyagaa ayuysaa
100101110011 HD23065
Inscrit en avril 2015

Our friends have put a like on the hypothesis of the NUMICON code copied and pasted and broken down into binary triples in Table 2.

The NUMICON code is linked to the identification of the star HD23065. It is indeed the star of the «Saliano».

This is most likely the good reason why our friends like it.

But it is not the result of an operation of Transposition of Table 1.

The like is in no way a validation of the response of the NUMICON code for Table 2. The NUMICON code has the same structure as the Binary Arrays, but that's another conundrum.

Contextual analysis of tables

In a first step, I would assume that the binary triples of the 12-square arrays are a binary coding of numbers in base 12 (numeration used by our visiting friends).

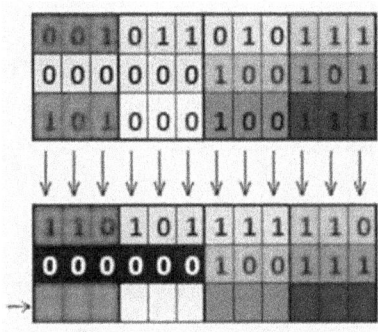

1 = ─	2 = Γ	3 = Π	4 = O	5 = D	6 = ○	7 = ○	8 = ○	9 = ○	10 = ○	11 = ○	12 = ≥

12 = ≥	21 = ─ꝺ	29 = Γꝺ	67 = ꝺꝺ	91 = ꝺ ꝺ
13 = ═	22 = ─ꝺ	30 = Γꝺ	68 = ꝺꝺ	93 = ꝺ ꝺ
14 = ─Γ	23 = ─ꝺ	31 = Γꝺ	75 = ꝺΠ	96 = ○≥
15 = ─Π	24 = Γ>	32 = Γ○	77 = ꝺꝺ	9⁸ = 9^8 = ○Γ
16 = ─O	25 = Γ─	33 = Γꝺ	80 = ꝺ○	100 = ○○
17 = ─ꝺ	26 = ΓΓ	34 = Γꝺ	82 = ꝺꝺ	101 = ○ꝺ
18 = ─ꝺ	27 = ΓΠ	35 = Γꝺ	85 = ꝺ─	105 = ○ꝺ
19 = ─ꝺ	28 = ΓO	36 = Π>		144 = ≥≥
20 = ─○				

And that each colored square encodes in binary a number in base 12 on 3 digits. The TRANSPOSITION to base 12 is therefore:

Table 1

0	0	12^0	0	12^1	12^0	0	12^1	0	12^2	12^1	12^0
0	0	0	0	0	0	12^2	0	0	12^2	0	12^0
12^2	0	12^0	0	0	0	12^2	0	0	12^2	12^1	12^0

Table 2

12^2	12^1	12^0	12^2	0	12^0	12^2	12^1	12^0	12^2	12^1	0
0	0	0	2	0	0	12^2	0	0	12^2	12^1	12^0

Analysis of Hypothesis 4: The Transposition of Value

For Value Transposition, any Base will give the same result.

The assumption of the transposition of the value is that Table 1 is the code issued for identification. It has a certain value.

Table 2 is the expected answer to validate the identification (Saliano in this case).

The expected answer is another encoding (given the 2 first lines) but which must give the initial value of Table 1.

This is the Transposition of Value

If so, it shows that the interlocutor has understood Table 1.

To make it easier for us to read, I give the values in Base 10. This gives us the Transposition :

Table 1

1, 13,12, 157 and total = 183

0, 0, 144, 145 and total = 289

145, 0, 144, 157 and total = 446

Total Value of Table 1 = 918

Table 2

156, 145, 157, 156 and total = 614

0.0, 144, 157 and total = 301

Total Table 2 Value = 915

So what is the value of the 3rd row of Table 2?

918-915=3

The answer is therefore 3 to be transposed to base 12 in the 4 boxes of the last row of table 2.

• To get the value 3 in this system, the only logical combination using simple integers is 1 + 1 + 1 + 0.

•That is to say three 001 blocks (Red/Value 1) and one 000 block (Black/Value 0).

• This solution uses exactly 3 «1» bits.

• Total final bits: 13 + 3 = 16. Perfect match.

The last row of Table 2 should consist of three 001 blocks and one 000 block (in any order, for example: 001,001,001,000). This ensures the Transposition of the Value (918 = 918) and the Transposition of the Elements (16 «1» bits retained).

In other words, 4 possible combinations.

Tableau 2, 4 solutions

1	1	0	1	0	1	1	1	1	1	1	0
0	0	0	0	0	0	1	0	0	1	1	1
0	0	0	0	0	1	0	0	1	0	0	1

1	1	0	1	0	1	1	1	1	1	1	0
0	0	0	0	0	0	1	0	0	1	1	1
0	0	1	0	0	0	0	0	1	0	0	1

1	1	0	1	0	1	1	1	1	1	1	0
0	0	0	0	0	0	1	0	0	1	1	1
0	0	1	0	0	1	0	0	0	0	0	1

1	1	0	1	0	1	1	1	1	1	1	0
0	0	0	0	0	0	1	0	0	1	1	1
0	0	1	0	0	1	0	0	1	0	0	0

Conclusion:

To verify the hypothesis of value transposition, we have four possible solutions.

Value transposition means that Table 1 is the code transmitted for identification.

Its value is 918 (base 10).

Table 2 is the expected response to validate the identification (of Saliano in this case).

The response is another encoding that yields the initial value of Table 1, 918 (Base 10).

The solution of «3 blocks of 001 and 1 block of 000» is the only one that allows to fill the delta of 3 in value and 3 in bits.

Value transposition demonstrates that the interlocutor has understood Table 1. According to our visiting friends, only ETs can do this.....

APPENDIX A: A TYPIST FROM THE UMMO FILE

Author: Denis Roger Denocla

With the kind collaboration of Jean-Jacques Pastor, José Juan Montejo Aguilera and Gema Lozano.

Table of Contents

INTRODUCTION

In 1982, a group of Spanish people investigated the UMMO case. They think that by discovering the typist who sends the mysterious letters, they will prove that these letters are a hoax.

Far from it, and quite the contrary. By identifying one of the typists, they will confirm the information contained in the UMMO file...

THE CONTEXT

Our visiting friends claim that the phalanges of their fingers are photosensitive. This would make them very sensitive to shocks. In fact, when they began their experiment in distributing typed documents in Spain in the mid-60s, typewriters required an undeniable power of typing with the fingers.

As a result, our visitors claim that they cannot use them and that they use the services of typists (the correct term is mechanographers, because there was no typing notation).

Our visitors claim to use the services of several people and alternate models of typewriters. This point has been noted on several occasions.

https://www.ummo-sciences.org/fr/D99.htm

The low neuroafferent activity of your epidermal sensory neurons in the hands surprised us, especially in the UNIOBIGAA (fingertips) and in the palms of the hands and wrists. We classified up to 37 cutaneous receptors, one of which is sensitive to mechanical frequencies up to 26,600 cycles/sec, others located in the outermost layers of the epidermis are sensitive to electromagnetic radiation corresponding to the 6.23 to 9.8 .104 c/s luminous and ultraviolet bands of 4.2 to 4.8 .1014. Thermal sensitivity is also more active. If we cannot distinguish with the wrists sharp images, on the other hand we can detect luminous surfaces (see note 31). The fingertips are particularly suitable for the integration of vibro-mechanical patterns because they are especially sensitive to these frequencies.

Note 31: With our wrist we can perceive vague colored surfaces, greenish, purple, purple, provided that the light intensity does not reach a certain threshold of blocking or inhibition, in this case, only the eyes continue to be sensitive to light stimulation. This allows us to orient ourselves in semi-darkness even with our eyes closed. Perception is monochrome regardless of the wavelength.

The palms of the hands are also photosensitive but to a lesser degree than the wrists. This prevents us from doing some work with your fingers that you can do. Strong percussion in the fingertips and wrists can cause serious damage to our sensory organs. Pressing very hard buttons, performing high-pressure gripping functions, typing on a machine are exercises where you have an advantage over us. Personally, I can testify to the real fears I had to endure when I arrived on Earth, in operations as harmless to you, as pressing certain buttons to operate elevators and electrical switches. When no Earthling sees me, I still use the knuckles of the fingers for these actions.

THE U3 SURVEY

```
LOCATION-FINDING ELEMENTS.
--------------------------------

1) The typist had placed an ad with address and phone number
   in the Madrid newspaper ABC.

2) Information extracted from his letter:

     - the Ummites opened his balcony wide (page 97).
     - There was a neon sign of an electricity and home
       appliances store just opposite his home. (page 98).
     - "They checked if there was anyone in the houses opposite
       on the other side of the street; these are not very far
       but not quite opposite his home. (page 98).
     - the gate only closed later. (page 98).
     - I learned that they were going to a hotel in the vicinity.
       (page I00).


INVESTIGATION CONDUCTED BY U 3?
--------------------------------

I) Research in the Madrid ABC between I January I960 and I
   July I967.
   A single name matches the ad with address and phone number:
   Mister FERNANDEZ 33 VIRIATO Street, Madrid.

2) On site:
     - the house matches the description in the document.
     - opposite there is indeed a store matching the information.
     - the balconies of the house opposite are indeed slightly offset.
     - there is indeed a gate at Mister Fernandez's house.
     - there is a hotel not very far from the house (hotel Trafalgar).

3) Other elements:
   -Mister Fernandez was a tenant of the apartment.
   -Villagrasa and Sesma lived in the neighborhood at that time.


4) Note that the owner of the apartment did not want to open to
the investigators, alleging late phone calls (3 in the morning)
regarding this Mr.cher Mr. Fernandez.
```

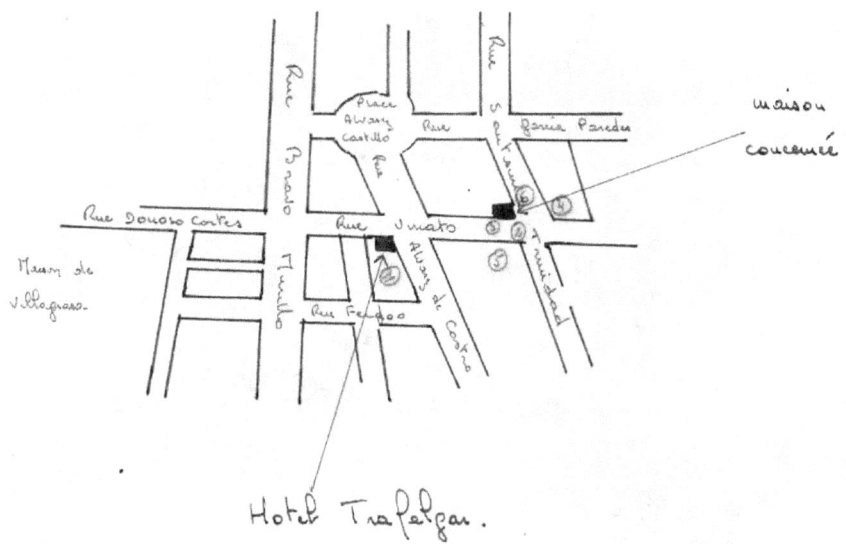

Rue Bretos Tirullo
Rue Donoso Cortes
Maison de Villagrasa.
Place Alvary Castillo
Rue
Rue Viriato
Alvars de Castro
Rue Feijoo
Rue S. de Trinidad
Garcia Paredes
Trinidad
maison concerné

Hotel Trafalgar.

Hotel Trafalgar, rue A. De Castro.

Façade dans l'angle Viriato/S. Trinidad.

215

Façade de la rue Viriato.

Édifice situé en face de la façade Rue González Trinidad.

Photo 4

. Édifice situé en face de la rue Viriato.

Photo 5

Magasin d'appareils électroménagers rue G. Trinidad.

Photo 6

COMPLEMENTS AND COMMENTS

Note that pages 97 to 100 mentioned in the U3 report must correspond to the book

«El misterio de Ummo», published by Antonio Ribera at Plaza & Janés in 1979.

While we don't have the details that led to the results presented, the U3 group seems to have taken things seriously. In fact, we will check the consistency and correlations between the results of the U3 group and the other elements at our disposal.

CONSISTENCY OF THE TYPIST'S CHRONOLOGY.

https://www.ummo-sciences.org/fr/data-E/E4.htm

The U3 group identified and located the mechanographer by identifying the advertisement that he had placed in the ABC newspaper between 1960 and 01/07/1967, but do not know the exact date of the advertisement.

He is Mr. Fernandez, a tenant at 33 Viriato Street in Madrid.

19/06/1967

«My wife and I, in the middle of moving these days, since we bought an apartment.»

At the very least, we can say that the announcement of the mechanographer was in ABC before 01/07/1967. This is compatible and consistent with the move of 19/06/1967, otherwise Mr. Fernandez would not have given this address in ABC.

Another element in favour of coherence is that our friends at UMMO have always been very generous with their partners. This could explain, at least partially, why the young couple can buy an apartment.

Finally, all the elements converge in a coherent way to estimate that Mr. Fernandez is young, i.e. a range between 20 and 25 years old.

GEOGRAPHICAL COHERENCE.

The typist lived in the same neighborhood where Mr. Enrique Villagrasa and Mr. Fernando Sesma also lived, who received letters from visitors to UMMO. A significant number of other recipients also resided in Madrid.

Fernando Sesma lived in Fernando el Católico Street (before 1966 he had published books, pamphlets and journalistic articles) and the second Enrique Villagrasa in Calle Donoso Cortés, two parallel streets perpendicular to Calle Bravo Murillo, located to the west of the latter. Between this street and the Argüelles district to Princesa Street.

CONFIRMATION OF SEVERAL TYPISTS

https://www.ummo-sciences.org/fr/data-E/E1.htm

(A priori Mr. Fernandez and his wife)...

moreover I know another person who, like me, writes their letters, a young typist

… administrative assistant...

And also and other mechanographer:

«which I didn't type myself because, lately, one more man was helping them write these things on the typewriter»

https://www.ummo-sciences.org/fr/data-E/E6.htm

There are therefore at least 3 mechanographers:

- M1: «perito mercantil», the salesman, Mr. Fernandez and his wife
- M2: The young administrative assistant
- M3: one more man

« ... They were seen by my concierge, from whom my wife had to seek an explanation to justify their coming to my house.»

(There is a translation error in this report in French, which speaks of «portal», but it is indeed «concierge».)

Although reserved on the subject, José Juan Montejo Aguilera did a little investigation, and verified that there was indeed a concierge at this address, named Petra. This is a complementary element for the verification of the consistency of the facts.

In other words, at M1's home when he wrote this letter to Sesma, and in any case before the move mentioned in his 1982 letter.

This move is mentioned in his letter of 19/06/1967. This means that, if the identification with Viriato-33 was correct, it must have been an address with a janitor's lodge at least in 1966/67.

IDENTIFICATION OF THE PROFESSOR OF MEDICINE

M1 wrote to «a professor of medicine»

https://www.ummo-sciences.org/fr/data-E/E9.htm

«Doctor of Medicine and for the moment, holder of a Chair at the Faculty of Madrid.»

Testimony of Gema Lozano:

«I had the honour of being a student of Professor Antonio Gallego Fernández, Professor of Special Physiology, in 1973, during my studies at the Faculty of Medicine of the Complutense University of Madrid. At that time, I was already familiar with the Ummo case, because in 1969 I had started attending the meetings of La Ballena Alegre until Mr. Sesma decided to close them. Unfortunately, when I was his student, I was unaware of the existence of this letter and the fact that he could have had this relationship with Ummo, and it was only years later that I learned about it.

It was said that the person could also be Professor Jorge Tamarit Torres, of whom I was also a student. He was a professor of general physiology and biochemistry and, at the time, rector of the faculty of medicine. I have no doubt that the author of this letter was Professor

Gallego, not only because of the way he expressed himself in this letter, but also because he was a true scholar in the specialty of neurophysiology, a disciple of Rafael Lorente de No, himself a disciple of Santiago Ramón y Cajal.

As for Professor Tamarit was an expert mainly in endocrinology and biochemistry (he also had a degree in physical sciences), but not so much in neurophysiology.

Therefore, and from the content of the letter, from what he talked about with the supposed Ummites, there is a 99% chance, in my opinion, that it is Professor Gallego; moreover, as far as I know, at that time he was perhaps the foremost expert in neurophysiology in Spain. Moreover, given the character of these two people (in my memory), I think that Professor Tamarit would never have written this letter; Besides, I think he wouldn't even have been willing to have multiple phone conversations with them.»

Gema Lozano contacted Professor Antonio Gallego Fernández by phone, who responded with information that was unknown to the mechanographer, Mr. Fernandez, and only mentioned in the letter E9.

No. 421

TELEPHONE CONVERSATION WITH DR. ANTONIO GALLEGO FERNANDEZ TODAY, 2 - 9 - 88

I called Dr. Antonio Gallego Fernandez, phone 91-2434498, which corresponds to his laboratory in the Faculty of Medicine, Physiology area, at the Complutense of Madrid.
    The Dr. ceased his activity as a professor of said faculty on September 30 of the year 1985. Currently, he has a research laboratory which he attends every day of the year, except holidays and feasts of obligation.

After my repetitive missives, 4 in total, and obtaining almost his agreement on the matter that concerns us, a great jar of cold water for the personnel!

I transcribe the conversation in the form that it developed.

Dr.--Hello?.
  --Good morning, I am the woman who has sent you the four letters in a period of three months.
Dr.--What do you want?.
  --It seemed interesting to me to speak with you regarding the last missive, on the third day of August, and I would like to reach an agreement to be able to visit you.
Dr.--Look, Miss, I know nothing about the matter and I have nothing to do with and I do not know what all that UMMO story is about.
  --Forgive me, I already understand that they have bothered you much by all this matter and I beg excuse if I have contributed to it, but you will understand that the matter is very interesting to many persons...
Dr.--Fine but for me it already has no importance and I beg that you leave me and do not send more letters. Regarding yours, I have shown them to all my colleagues and they are unaware of the existence of that apparatuss and of the camera with which the film.
  -- Sir, please think that I have promised you discretion on my part and....
Dr.--!miss,let us leave it be!, Agreement?. !Goodbye!
  --!Excuse me!...

Without another word, he hung up the telephone on me, leaving my ear a total mess.

<u>CONTRADICTIONS</u>
--....But for me it already has no importance. For the heard, I believe that at some time yes it had it, since that 'Already' affirms it.

<u>Film= Movie.</u>
If you review my letter dated June 27 of the year in course, this one says in one of its last paragraphs.
"I also seem to remember that you took some photos of the mechanism with a KODAC camera and that two days later a gentleman with a common trait, he possessed a beard and of young age 25-30 years, went to pick up the apparatus." (Possibly the typists.
In this conversation with the Professor, he affirms to me that a film of the apparatus was taken, since I in no moment has commented on it. My letter cited above says photos, not film or Movie.

In this short conversation he was noticeable nervous, besides his reiterated desire to hang up the phone on me. I have experience in this type of thing, since two years ago, on August 26, 1986, the Civil Governor of Albacete, he threw me almost with kicks from the Civil Government of that locality when I wanted to deepen an investigation about Sanitation that brought me to that city of the Lord.

Anyway, my conversation with Dr. Gallego and my visits to Albacete have not
concluded.

Regarding the identity of the other "North American cardiologist" doctor, to whom
the Ummis refer in one of their communications, it seems with all certainty to be:

Mr Thomas Spriengfield
Hoster,R.M.
------ (U.S.A.)

He received information from the inhabitants of UMMO in 1.966-68 and participated
for some years in a mediine project for astronauts at NASA in Pasadena.

Confidential information for RIBERA-FARRIOLS- DARNAUDE

SIGNED: GEMA LOZANO

In conclusion

The Mechanographer

The elements gathered allow us to establish with a high probability that several mechanographers operated on behalf of the visitors of UMMO, and that one of them was Mr. Fernandez, tenant at 33 Viriato Street in Madrid.

If Mr. Fernandez was in his twenties at the time, in 2025, he would be about 80-85 years old.

The Professor

The elements gathered allow us to establish with a high probability that Professor Antonio Gallego Fernández, was indeed related to the mechanographer M. Fernandez.

THE MYSTERIOUS PEBBLE OF FERNANDO SESMA

Writing: Denis R. Denocla

Photo: JJ Pastor

Working Group: D. M. Fistroe, Banban, JJ Pastor, J. J. Montejo, Denis R. Denocla

Contents

CONTEXT OF THE PHOTO OF THE OBJECT

The object comes from the Sesma collection purchased by Rafael Farriols from Fernando Sesma in 1972.

The photo of the object was taken by JJ Pastor at Rafael Farriols' house in 1973 during the Barcelona meeting.

This mysterious engraved rock has been without serious explanation for more than 50 years.

Reference to the object in the corpus U.

D1378 / Ignacio Darnaude, UMMOCAT (N 3 467, N 3 577, N 3 819)

To: Javier Serra and/or Barrenechea

Date: 30/01/1988 confidential

Date: public release 1996

Fernando Sesma greeted the first phone calls with a certain degree of skepticism. We gathered a great deal of data on his neuromental structure through his phonation, and we realized that his intellectual-emotional models could be positively influenced by giving him a message engraved on the surface of a stone.

Thus began the first telephone conversations with your unfortunate brother (in 1965), who later fell victim to the despicable manipulations of other intragalactic beings.

ITEM DESCRIPTION

My first impression was that the object was made of glazed pottery, JJ Pastor confirmed that it was indeed a pebble. The brown color on which the symbols are engraved seemed to JJ Pastor to be a kind of natural gangue. Given the shape and surface appearance of the brown stain, we cannot rule out a varnish, probably an in situ examination would make it possible to decide.

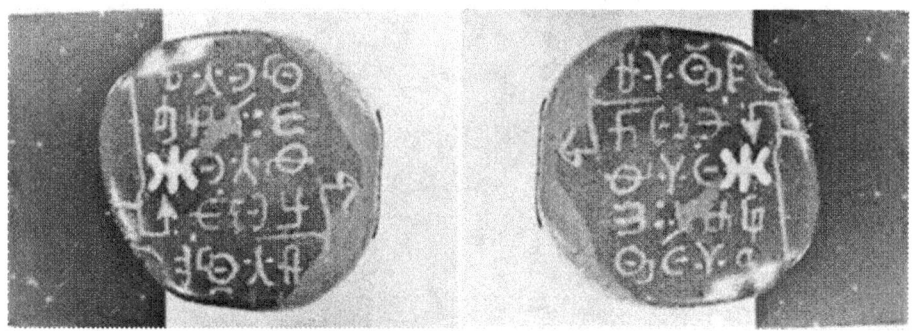

Left View — Right View

OBJECT ANALYSIS

The pebble is relatively ovoid, its diameter is about 10-12 cm.

The pebble has symbols engraved on one side only, most of which is on the brown area.

PAPER DRAWING

The engraved pebble was accompanied by a drawing on paper of the ideograms.

This drawing is also an original.

DESCRIPTION OF THE SYMBOLS

It looks like there are 20 symbols and 2 arrows. The symbols are relatively well aligned horizontally and vertically, in 5 lines of 4 symbols. One of the symbols is engraved in «bold»

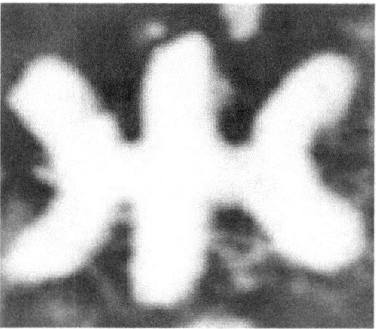

SYMBOL ANALYSIS

a)   A first hypothesis would be Berber symbols. In this case, the Berber symbol in bold should be read vertically.

The Berbers used different alphabets, but no other symbol corresponds to them in any significant way.

The hypothesis of Berber symbols can be ruled out without a doubt.

b)   The hypothesis of U. we recognize with high probability the ideogram

«UMMOAELEWEE GENERAL COUNCIL OF UMMO»

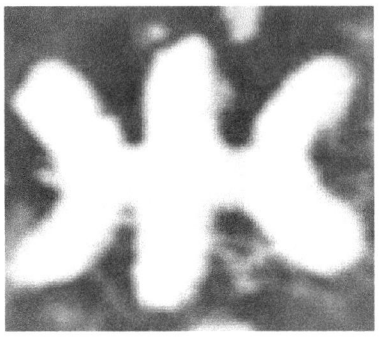 

Sesma Stone   S68-s1

UMMOAELEWEE

UMMO GENERAL COUNCIL

b.1)   In this case, only the Left or Right views are possible, given the bold U symbol.

Left View — Right View

b.2)   In the right view, 1st line, 4th ideogram, we recognize with a high probability the ideogram

«HERE/THIS PLACE»

SIMARII

Concept «this place» SIMARII = [S] cycle «a» [[I] identification «a' [[M] join «a' [[A] displacement 'a' [R] superapposition 'a' [II] limit

[S] cycle 'a' [[I] identification 'a' [[M] join 'a' [[A] displacement 'a' delimited overapposition]

[S] cycle 'a' [[I] identification 'a' [[M] join 'a' [displacement has a delimited overapposition] [S] cycle 'a' [[I] identification attached to the displacement of a delimited overapposition identified cycle attached to the displacement of a delimited overapposition]

Proposed translation: Perimeter/contour joined to the surface [inner perimeter/contour] In summary: this area is delimited

PRESENCE 6 — The LANGUAGE of the People of UMMO DENOCLA DICTIONARY

Sesma's Pebbles   OT 20

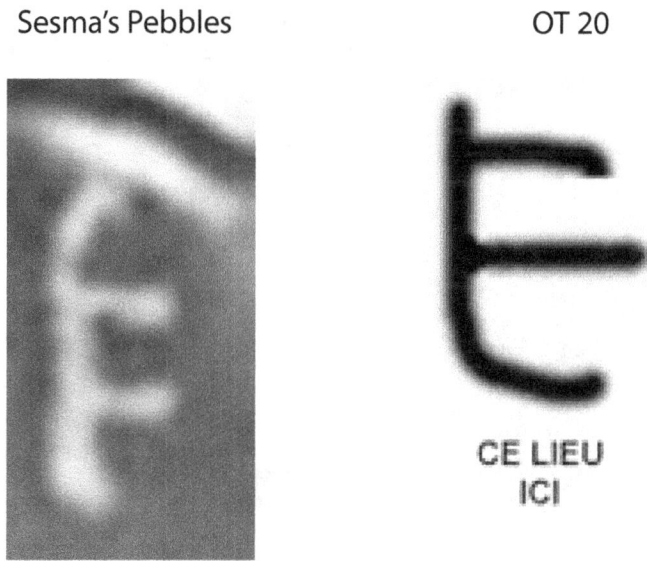

CE LIEU
ICI

b.3) In right view, 1st line, 2nd ideogram,

In right view, 3rd line, 2nd ideogram, In right view, 5th line, 3rd ideogram,

we recognize with medium probability, given the absence of lateral points, the ideogram

'PURPOSE'

Sesma's Pebbles   OT 20

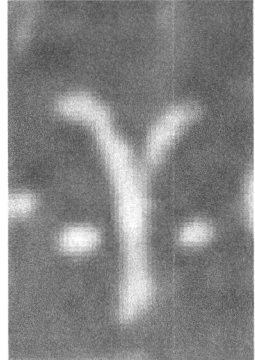

OBJET

b.4.1) In right view, 5th line, 4th ideogram, on the photo of the pebble, we recognize with a high probability, the ideogram of the number/digit '7'

Sesma's Pebbles   D45

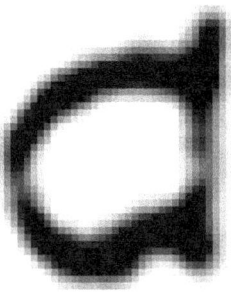

b.4.2) In right view, 5th line, 4th ideogram, on the drawing,

The symbol is markedly different from the number 7 and is unknown.

b.5) The other symbols are unknown.

DISCUSSION OF SYMBOLS

A) Our first feeling

Our first feeling of the symbols of the Sesma pebble is suspicion. This is for 2 reasons:

- The UMMO symbol is in bold, obviously to draw attention to it, which seems suspicious to us

- The 2 arrows are engraved in the traditional terrestrial way, while the people of UMMO use 'hook' arrows.

However, if we consider that the symbol of UMMO is in bold, obviously to attract the attention of Fernando Sesma, as indicated in document D1378, then the symbol of UMMO in bold is not suspect. On the contrary, it is consistent and lends credence to the authenticity of Sesma's pebble.

Just as the documents were typed on a typewriter by mundane, it is reasonable to think that the symbols were also engraved by a mundane. This would explain the land-style spires.

B) The results of the ideogram analysis

• 2 ideograms out of 20 are identifiable with very high probability

- 1 ideogram out of 20 is identifiable with medium probability

- 1 in 20 ideograms is identifiable with poor probability

The result of 10% of the identification of the ideograms is quantitatively low, but qualitatively with very high probability for these 2 ideograms. This result also validates the reading of the pebble in Right View.

C) Hypothesis on a possible decipherment

As a general rule, 1 or 2 ideograms can be associated with a 'word' or a feature.

There are very few ideograms of common words known in corpus U. This makes this possible deciphering very difficult. The deciphering hypothesis does not impact the chronological analysis.

We have 5 lines of 4 ideograms. Maybe 5 'sentences' or 'propositions' of 3 or 4 'words'.

1. On the first line, we would have: X 'object' Y 'here'

In the present context, we could imagine a sentence like:

a. (ideo) (grams) engraved here

b. The pebble engraved here

2. On the 2nd line, we might have something like:

'for you Mr. Sesma'

3. On the 3rd line, we might have something like:

X 'subject' Y 'general council of UMMO'

In the present context, we could imagine a sentence like:

a. (ideo) (grams) given by UMMO

b. the pebble donated by UMMO

4. On the 4th line, we might have something like:

In the present context, we could imagine a sentence like:

'With respectful regards'

5. On the 5th line, we might have something like: X Y 'object' (7 or Z)

In the present context, we could imagine a sentence like:

a. X (ideo) (grams) (7 or Z) => reference U. of the object for internal tracking U. ?

b. X the pebble (7 or Z) => reference U. of the object for internal tracking U. ?

In summary, I will favor the hypothesis of the following translation:

The ideograms engraved here are for you Mr. Sesma,

The ideograms are a gift from UMMO, with our respectful greetings,

(object reference)

DISCUSSION OF THE TIMELINE

Long before this letter was written, at the end of 1954 when Fernando Sesma created the Sociedad de Amigos de los Visitantes del Espacio. The group met at the Ballena Alegre and 2 weeks after the creation of the association, Alberto Sanmartín Comes declared that he had been contacted, and had a stone on which a message was written. Sesma invites him.

The Alberto Sanmartín Comes rock has 9 symbols, while the Alberto Sanmartín Comes stone has 20 symbols. So they are not the same.

Ignacio Darnaude, UMMOCAT, N° 4,335—The violet stone tablet from space. "… two of the nine symbols engraved on the said rock...»

It is certain that the creation of the tablet by Alberto Sanmartín Comes, a priori 1954, is much earlier than the letter D1378 of 30-1-1988. This 9-symbol rock cannot therefore be a fraud inspired by the D1378. Alberto Sanmartín Comes is not one of the well-known people put under mind control by the Salianos. Given the changing versions of the story, the probability of fraud, inspired by the Adamski affair of the current events of the time, is very high for Alberto Sanmartín Comes' tablet .

Letter D1378 of 30-1-1988 to J. Barrenechea and his wife Carmen Maria, remained confidential until 1996. It mentions a message engraved on a pebble by our friends U. for Fernando Sesma. The U. began by contacting Fernando Sesma by phone in 1965 and gave him a pebble with an engraved message and a paper document of the same message, probably quickly, that same year 1965.

Thus, the irrational and whimsical Fernando Sesma would be more receptive to their telephone communications... Fernando Sesma is said to have been in possession of the U. rock from 1965 to 1973, when it was bought by Rafael Farriols and photographed by JJ Pastor. Here again, it is certain that the construction of the Fernando Sesma stone predates the letter D1378 of 30-1-1988. This pebble with 20 symbols cannot therefore be a fraud inspired by the D1378.

1954	1965	1972	1988	1996
Pebble	Pebble	Pebble	Letter writing	Circulation letter
Alberto Sanmartín Comes	Fernando	Fernando	D1378	D1378
	Sesma	Sesma		
		bought by Rafael Farriols		
		(Photo)		

CONCLUSION

Given the chronology, the 2 pebbles cannot therefore be a fraud inspired by the D1378.

Given that the letter D1378 is a validated original, a pebble with the ideogram U. was indeed given by our friends to Sesma in 1965.

Given the analysis of the symbols of the pebble held by Rafael Farriols and photographed by JJP, and that Alberto Sanmartín Comes' tablet is distinct from Fernando Sesma's pebble, it is certain that this is indeed the pebble given by our visitors U. to Fernando Sesma in 1965.

This mysterious engraved stone had remained without serious explanation for more than 50 years, let's hope that this object joins collections for the heritage of humanity.

MANIFESTO FOR THE EXOCIVILIZATION'S RECOGNITION

This manifesto outlines a few of the basic principles necessary to establish a fair and long-lasting relationship with any exocivilization respecting the *"Pax Galactica"*:

Exocivilizations' Rights

. Official recognition of the exocivilizations
. Application of Human Rights to all exocivilizations
. Application of the Geneva Convention to all exocivilizations
. Restitution of the bodies of dead explorers to their exocivilizations
(in reference to the 1994 Law on Bioethics)

Exocivilizations' Duties

. Respect for the UN conventions and resolutions
. Respect for the rights of States
. Respect for the rights and integrity of people and property

D. R. DENOCLA

"Knowledge for whom? Knowledge for what?"

REMEMBER TO POST
YOUR COMMENT ON AMAZON

13

BIBLIOGRAPHY

The sources of Ummo documents are the site http://www.ummo-sciences.org and http://www.ummo-ciencias.org and D.R. DENOCLA

« Le langage des Ummites : du chinois ? », Johannes Gehrs, Inforespace n° 103, décembre 2001, pp. 39-55.

« UmmoCat : Documental Catalogue of the Cryptic Group Ummo », Ignacio Darnaude, édition privée, 1982 –2001, 4 volumes , 1280 p.

« Genèses : l'Univers, le Vivant, l'Homme », Notes de recherche Tome 3, 2003-2005. Edition privée sur le site http://www.denocla.com, 100 p.

« La tétravalence expliquée aux enfants », Frédéric Morin, Morpheus n° 11 septembre 2005, page 8.

Alban Nanty http://www.ummo-sciences.org/activ/art/art8.htm

Alfred North Whitehead et Bertrand Russell, « Principia Mathematica » 1913.

Bertrand Russell, « Philosophie de l'atomisme logique », 1918.

Gottlob Frege « une écriture conceptuelle » 1879.

http://fr.wikipedia.org/wiki/Bertrand_Russell#Logique

« LA PHILOSOPHIE DE LA LOGIQUE », Précis de Philosophie analytique, Michel Seymour, Éditions du Seuil - http://www.philo.umontreal.ca/textes/Seymour_LOGIQUE.pdf

Jean Pollion, « Ummo des vrais extraterrestres ! », Edition Aldane, 2002.

Ludwig Wittgenstein « Tractatus logico-philosophicus », suivi de « Investigations philosophiques », trad. de Pierre Klossowski, Paris, Gallimard, 1961.

Michel Seymour « La Philosophie de la Logique »

Multiples documents et analyses sur le site http://www.ummo-sciences.org

Norman Molhant http://www.cafe.edu/sf/pl4c/

www.ummo-sciences.org,multiplesdocumentsetanalysessurlesite.

www.cafe.edu/sf/pl4c, Norman Molhant.

Référencessurlogiquetétravalente,www.ummo-sciences.org/activ/science/tetra/index.htm

Jacques Pazelle, communications personnelles.

Aux frontières de Wolf 424 le soleil de Ummo? Alain Ranguis www.ummo-sciences.org/activ/art/art2.htm

D.R.Denocla/JacquesPazelle,puisManuelRotaechecommunications Pluton — 2003 UB313 — D116.

Le système de numération des Oummains — D.R. Denocla.

Étude commentée du moteur à plasma, Jean Pollion 10-2003, www.ummo-sciences.org/activ/analyses/ana14.htm

Vicenç Sole i Ferré communication du 11 juin 2007, Erreurs annexe Alicia Araujo.

Communications«mortalitéetlongévité»,ManuelRotaeche,André-JacquesHolbec,www.ummo-sciences.org/activ/ex-adummo/debat6.htm

Bibliographie relative au Germanium :
www.societechimiquedefrance.fr/produit-du-jour/germanium.html
www.sciencedirect.com/science/article/pii/002236976190141X
http://fr.wikipedia.org/wiki/Germanium
http://en.wikipedia.org/wiki/Fiber-optic
http://en.wikipedia.org/wiki/Germanium_dioxide
www.mineralinfo.org/Substance/Germanium/GeDCE.pdf

© 2012, UMMO WORLD Publishing
8 Esp. de la Manufacture
92136 Issy-les-Moulineaux

Imprimé par :
Graphic Systems.Com
69 chemin de la Chapelle St Antoine
95300 Ennery

Achevé d'imprimer en septembre 2012
Dépôt légal : septembre 2012
Imprimé en France

www.ingramcontent.com/pod-product-compliance
Lightning Source LLC
Chambersburg PA
CBHW080802180526

45168CB00006B/2304